When Ei...
Walked with Gödel

"Jim Holt's essay collection *When Einstein Walked with Gödel* is a gleaming introduction to the mysteries of modern physics and mathematics. Holding your hand through discussions of artificial intelligence, string theory, theories of time, Holt's 'Excursions to the Edge of Thought' are the journeys of a lifetime."
—Christopher Bray, *The Tablet* (London)

"Science writing of the caliber on display in *When Einstein Walked with Gödel* is a boon in these times of looming scientific illiteracy. Holt makes his recondite subjects seem not only fascinating but fun."
—Steve Donoghue, *The Christian Science Monitor*

"A smart, erudite, and witty guided tour of some of the most colorful episodes and characters in the history of science and mathematics."
—Jimena Canales, *Undark*

"I've just discovered the brilliant essays of Jim Holt . . . *When Einstein Walked with Gödel: Excursions to the Edge of Thought* . . . offers lucid and entertaining accounts of deep problems in physics, maths and philosophy."
—Martin Rees, *Physics World*

"These are bold, thought-provoking pieces . . . These are stories of real humans and their mathematical, physical and philosophical theories—some of the most complex ever devised."
—Andrew Jaffe, *Nature*

"[A] fantastic essay collection [filled] with stories about eccentric geniuses and groundbreaking ideas at the intersection of science and philosophy . . . Holt delivers this feast of wild genius, oddball thinkers, and sheer creativity in his signature accessible style of writing and [a] playful tone."
—*Publishers Weekly* (starred review)

"Jim Holt's *When Einstein Walked with Gödel* is a thrilling trek through some of the greatest insights in physics, philosophy, and mathematics. Insightful, enlightening, and entertaining, Holt explores how a collection of thinkers—some quirky, some tragic, all ingenious—redefined the very boundaries of space, time, and knowledge."

—Brian Greene, author of *The Hidden Reality: Parallel Universes and the Deep Laws of the Cosmos*

"Jim Holt's essays are full of wonder and wisdom, irreverence and wit. And they also possess a special quality: reading them is like getting a joke—beyond the words, there's a sense of revelation and unity, ours to enjoy."

—Edward Frenkel, Professor of Mathematics at the University of California, Berkeley, and author of *Love & Math: The Heart of Hidden Reality*

"Jim Holt not only has an unerring sense for locating the most interesting questions lying on the borders of philosophy, science, and mathematics; he also has a talent for expressing the human and emotional dimensions of the life of the mind. The blend goes toward making *When Einstein Walked with Gödel* an unusually absorbing and stimulating collection of essays."

—Rebecca Newberger Goldstein, author of *Plato at the Googleplex: Why Philosophy Won't Go Away*

JIM HOLT

When Einstein
Walked with Gödel

Jim Holt writes about math, science, and philosophy for *The New York Times*, *The New Yorker*, *The Wall Street Journal*, and *The New York Review of Books*. His book *Why Does the World Exist? An Existential Detective Story* was an international bestseller.

When Einstein
Walked with Gödel

When Einstein Walked with Gödel

EXCURSIONS TO THE EDGE OF THOUGHT

Jim Holt

FARRAR, STRAUS AND GIROUX   NEW YORK

Farrar, Straus and Giroux
175 Varick Street, New York 10014

Printed in the United States of America
Published in 2018 by Farrar, Straus and Giroux
First paperback edition, 2019

The Library of Congress has cataloged the hardcover edition as follows:
Names: Holt, Jim, 1954– author.
Title: When Einstein walked with Gödel : excursions to the edge of thought /
Jim Holt.
Description: First edition. | New York : Farrar, Straus and Giroux, 2018. |
Includes index.
Identifiers: LCCN 2017038334 | ISBN 9780374146702 (hardcover) |
ISBN 9780374717841 (ebook)
Classification: LCC PS3608.O4943595 A6 2018 | DDC 814/.6—dc23
LC record available at https://lccn.loc.gov/2017038334

Paperback ISBN: 978-0-374-53842-2

Designed by Jonathan D. Lippincott

Our books may be purchased in bulk for promotional, educational, or business use.
Please contact your local bookseller or the Macmillan Corporate and
Premium Sales Department at 1-800-221-7945, extension 5442, or by e-mail at
MacmillanSpecialMarkets@macmillan.com.

www.fsgbooks.com
www.twitter.com/fsgbooks • www.facebook.com/fsgbooks

1 3 5 7 9 10 8 6 4 2

To the memory of Bob Silvers

Contents

Preface

These essays were written over the last two decades. I selected them with three considerations in mind.

First, the depth, power, and sheer beauty of the ideas they convey. Einstein's theory of relativity (both special and general), quantum mechanics, group theory, infinity and the infinitesimal, Turing's theory of computability and the "decision problem," Gödel's incompleteness theorems, prime numbers and the Riemann zeta conjecture, category theory, topology, higher dimensions, fractals, statistical regression and the "bell curve," the theory of truth—these are among the most thrilling (and humbling) intellectual achievements I've encountered in my life. All are explained in the course of these essays. My ideal is the cocktail-party chat: getting across a profound idea in a brisk and amusing way to an interested friend by stripping it down to its essence (perhaps with a few swift pencil strokes on a napkin). The goal is to enlighten the newcomer while providing a novel twist that will please the expert. And never to bore.

My second consideration is the human factor. All these ideas come with flesh-and-blood progenitors who led highly dramatic lives. Often these lives contain an element of absurdity. The creator of modern statistics (and originator of the phrase "nature versus nurture"), Sir Francis Galton, was a Victorian prig who had comical misadventures in the African bush. A central figure in the history of the "four-color theorem" was a flamboyantly eccentric mathematician/classicist called Percy

Heawood—or "Pussy" Heawood by his friends, because of his feline whiskers.

More often the life has a tragic arc. The originator of group theory, Évariste Galois, was killed in a duel before he reached his twenty-first birthday. The most revolutionary mathematician of the last half century, Alexander Grothendieck, ended his turbulent days as a delusional hermit in the Pyrenees. The creator of the theory of infinity, Georg Cantor, was a kabbalistic mystic who died in an insane asylum. Ada Lovelace, the cult goddess of cyber feminism (and namesake of the programming language used by the U.S. Department of Defense), was plagued by nervous crises brought on by her obsession with atoning for the incestuous excesses of her father, Lord Byron. The great Russian masters of infinity, Dmitri Egorov and Pavel Florensky, were denounced for their antimaterialist spiritualism and murdered in Stalin's Gulag. Kurt Gödel, the greatest of all modern logicians, starved himself to death out of the paranoiac belief that there was a universal conspiracy to poison him. David Foster Wallace (whose attempt to grapple with the subject of infinity I examine) hanged himself. And Alan Turing—who conceived of the computer, solved the greatest logic problem of his time, and saved countless lives by cracking the Nazi "Enigma" code—took his own life, for reasons that remain mysterious, by biting into a cyanide-laced apple.

My third consideration in bringing these essays together is a philosophical one. The ideas they present all bear crucially on our most general conception of the world (metaphysics), on how we come to attain and justify our knowledge (epistemology), and even on how we conduct our lives (ethics).

Start with metaphysics. The idea of the infinitely small—the infinitesimal—raises the question of whether reality is more like a barrel of molasses (continuous) or a heap of sand (discrete). Einstein's relativity theory either challenges our notion of time or—if Gödel's ingenious reasoning is to be credited—abolishes it altogether. Quantum entanglement calls the reality of space into question, raising the possibility that we live in a "holistic" universe. Turing's theory of computability forces us to rethink how mind and consciousness arise from matter.

Then there's epistemology. Most great mathematicians claim insight into an eternal realm of abstract forms transcending the ordinary world we live in. How do they interact with this supposed "Platonic" world to obtain mathematical knowledge? Or could it be that they are radically mistaken—that mathematics, for all its power and utility, ultimately amounts to a mere tautology, like the proposition "A brown cow is a cow"? To make this issue vivid, I approach it in a novel way, by considering what is universally acknowledged to be the greatest unsolved problem in mathematics: the Riemann zeta conjecture.

Physicists, too, are prone to a romantic image of how they arrive at knowledge. When they don't have hard experimental/observational evidence to go on, they rely on their aesthetic intuition—on what the Nobel laureate Steven Weinberg unblushingly calls their "sense of beauty." The "beauty=truth" equation has served physicists well for much of the last century. But—as I ask in my essay "The String Theory Wars"— has it recently been leading them astray?

Finally, ethics. These essays touch on the conduct of life in many ways. The eugenic programs in Europe and the United States ushered in by the theoretical speculation of Sir Francis Galton cruelly illustrate how science can pervert ethics. The ongoing transformation of our habits of life by the computer should move us to think hard about the nature of happiness and creative fulfillment (as I do in "Smarter, Happier, More Productive"). And the omnipresence of suffering in the world should make us wonder what limits there are, if any, to the demands that morality imposes upon us (as I do in "On Moral Sainthood").

The last essay in the volume, "Say Anything," begins by examining Harry Frankfurt's famous characterization of the bullshitter as one who is not hostile to the truth but indifferent to it. It then enlarges the picture by considering how philosophers have talked about truth— erroneously?—as a "correspondence" between language and the world. In a slightly ludic way, this essay bridges the fields of metaphysics, epistemology, and ethics, lending the volume a unity that I hope is not wholly specious.

And lest I be accused of inconsistency, let me (overconfidently?) express the conviction that the "Copernican principle," "Gödel's incompleteness theorems," "Heisenberg's uncertainty principle," "Newcomb's problem," and "the Monty Hall problem" are all exceptions to Stigler's law of eponymy (*vide* p. 292).

J.H.
New York City, 2017

PART I

The Moving Image
of Eternity

When Einstein Walked with Gödel

In 1933, with his great scientific discoveries behind him, Albert Einstein came to America. He spent the last twenty-two years of his life in Princeton, New Jersey, where he had been recruited as the star member of the Institute for Advanced Study. Einstein was reasonably content with his new milieu, taking its pretensions in stride. "Princeton is a wonderful piece of earth, and at the same time an exceedingly amusing ceremonial backwater of tiny spindle-shanked demigods," he observed. His daily routine began with a leisurely walk from his house, at 112 Mercer Street, to his office at the institute. He was by then one of the most famous and, with his distinctive appearance—the whirl of pillow-combed hair, the baggy pants held up by suspenders—most recognizable people in the world.

A decade after arriving in Princeton, Einstein acquired a walking companion, a much younger man who, next to the rumpled Einstein, cut a dapper figure in a white linen suit and matching fedora. The two would talk animatedly in German on their morning amble to the institute and again, later in the day, on their way homeward. The man in the suit might not have been recognized by many townspeople, but Einstein addressed him as a peer, someone who, like him, had single-handedly launched a conceptual revolution. If Einstein had upended our everyday notions about the physical world with his theory of relativity, the younger man, Kurt Gödel, had had a similarly subversive effect on our understanding of the abstract world of mathematics.

Gödel, who has often been called the greatest logician since Aristotle, was a strange and ultimately tragic man. Whereas Einstein was gregarious and full of laughter, Gödel was solemn, solitary, and pessimistic. Einstein, a passionate amateur violinist, loved Beethoven and Mozart. Gödel's taste ran in another direction: his favorite movie was Walt Disney's *Snow White and the Seven Dwarfs*, and when his wife put a pink flamingo in their front yard, he pronounced it *furchtbar herzig*— "awfully charming." Einstein freely indulged his appetite for heavy German cooking; Gödel subsisted on a valetudinarian's diet of butter, baby food, and laxatives. Although Einstein's private life was not without its complications, outwardly he was jolly and at home in the world. Gödel, by contrast, had a tendency toward paranoia. He believed in ghosts; he had a morbid dread of being poisoned by refrigerator gases; he refused to go out when certain distinguished mathematicians were in town, apparently out of concern that they might try to kill him. "Every chaos is a wrong appearance," he insisted—the paranoiac's first axiom.

Although other members of the institute found the gloomy logician baffling and unapproachable, Einstein told people that he went to his office "just to have the privilege of walking home with Kurt Gödel." Part of the reason, it seems, was that Gödel was undaunted by Einstein's reputation and did not hesitate to challenge his ideas. As another member of the institute, the physicist Freeman Dyson, observed, "Gödel was . . . the only one of our colleagues who walked and talked on equal terms with Einstein." But if Einstein and Gödel seemed to exist on a higher plane than the rest of humanity, it was also true that they had become, in Einstein's words, "museum pieces." Einstein never accepted the quantum theory of Niels Bohr and Werner Heisenberg. Gödel believed that mathematical abstractions were every bit as real as tables and chairs, a view that philosophers had come to regard as laughably naive. Both Gödel and Einstein insisted that the world is independent of our minds yet rationally organized and open to human understanding. United by a shared sense of intellectual isolation, they found solace in their companionship. "They didn't want to speak to anybody else," another member of the institute said. "They only wanted to speak to each other."

People wondered what they spoke about. Politics was presumably one theme. (Einstein, who supported Adlai Stevenson, was exasperated when Gödel chose to vote for Dwight D. Eisenhower in 1952.) Physics was no doubt another. Gödel was well versed in the subject; he shared Einstein's mistrust of the quantum theory, but he was also skeptical of the older physicist's ambition to supersede it with a "unified field theory" that would encompass all known forces in a deterministic framework. Both were attracted to problems that were, in Einstein's words, of "genuine importance," problems pertaining to the most basic elements of reality. Gödel was especially preoccupied by the nature of time, which, he told a friend, was *the* philosophical question. How could such a "mysterious and seemingly self-contradictory" thing, he wondered, "form the basis of the world's and our own existence"? That was a matter in which Einstein had shown some expertise.

Decades before, in 1905, Einstein proved that time, as it had been understood by scientist and layman alike, was a fiction. And this was scarcely his only achievement that year. As it began, Einstein, twenty-five years old, was employed as an inspector in a patent office in Bern, Switzerland. Having earlier failed to get his doctorate in physics, he had temporarily given up on the idea of an academic career, telling a friend that "the whole comedy has become boring." He had recently read a book by Henri Poincaré, a French mathematician of enormous reputation, that identified three fundamental unsolved problems in science. The first concerned the "photoelectric effect": How did ultraviolet light knock electrons off the surface of a piece of metal? The second concerned "Brownian motion": Why did pollen particles suspended in water move about in a random zigzag pattern? The third concerned the "luminiferous ether" that was supposed to fill all of space and serve as the medium through which light waves moved, the way sound waves move through air, or ocean waves through water: Why had experiments failed to detect the earth's motion through this ether?

Each of these problems had the potential to reveal what Einstein held to be the underlying simplicity of nature. Working alone, apart from the scientific community, the unknown junior clerk rapidly managed to dispatch all three. His solutions were presented in four papers, written in March, April, May, and June of 1905. In his March paper, on the photoelectric effect, he deduced that light came in discrete particles, which were later dubbed photons. In his April and May papers, he established once and for all the reality of atoms, giving a theoretical estimate of their size and showing how their bumping around caused Brownian motion. In his June paper, on the ether problem, he unveiled his theory of relativity. Then, as a sort of encore, he published a three-page note in September containing the most famous equation of all time: $E = mc^2$.

All these papers had a touch of magic about them and upset some deeply held convictions in the physics community. Yet, for scope and audacity, Einstein's June paper stood out. In thirty succinct pages, he completely rewrote the laws of physics. He began with two stark principles. First, the laws of physics are absolute: the same laws must be valid for all observers. Second, the speed of light is absolute; it, too, is the same for all observers. The second principle, though less obvious, had the same sort of logic to recommend it. Because light is an electromagnetic wave (this had been known since the nineteenth century), its speed is fixed by the laws of electromagnetism; those laws ought to be the same for all observers; and therefore everyone should see light moving at the same speed, regardless of their frame of reference. Still, it was bold of Einstein to embrace the light principle, for its consequences seemed downright absurd.

Suppose—to make things vivid—that the speed of light is a hundred miles an hour. Now suppose I am standing by the side of the road and I see a light beam pass by at this speed. Then I see you chasing after it in a car at sixty miles an hour. To me, it appears that the light beam is outpacing you by forty miles an hour. But you, from inside your car, must see the beam escaping you at a hundred miles an hour, just as you would if you were standing still: that is what the light principle demands. What if you gun your engine and speed up to ninety-nine miles an hour? Now I see the beam of light outpacing you by just one mile an hour. Yet to you, inside the car, the beam is still racing ahead at a hundred miles an hour, despite your increased speed. How can this be? Speed, of course, equals distance divided by time. Evidently, the faster you go in your car, the shorter your ruler must become and the slower your clock must tick relative to mine; that is the only way we can continue to agree on the speed of light. (If I were to pull out a pair of binoculars and look at your speeding car, I would actually see its length contracted and you moving in slow motion inside.) So Einstein set about recasting the laws of physics accordingly. To make these laws absolute, he made distance and time relative.

It was the sacrifice of absolute time that was most stunning. Isaac Newton believed that time was objective, universal, and transcendent of all natural phenomena; "the flowing of absolute time is not liable to any change," he declared at the beginning of his *Principia*. Einstein, however, realized that our idea of time is something we abstract from our experience with rhythmic phenomena: heartbeats, planetary rotations and revolutions, the ticking of clocks. Time judgments always come down to judgments of simultaneity. "If, for instance, I say, 'That train arrives here at 7 o'clock,' I mean something like this: 'The pointing of the small hand of my watch to 7 and the arrival of the train are simultaneous events,'" Einstein wrote in the June paper. If the events in question are at some distance from each other, judgments of simultaneity can be made only by sending light signals back and forth. Working from his two basic principles, Einstein proved that whether an observer deems two events to be happening "at the same time" depends on his state of motion. In other words, there is no universal *now*. With different observers slicing up the timescape into "past," "present," and "future" in different ways, it seems to follow that all moments coexist with equal reality.

Einstein's conclusions were the product of pure thought, proceeding from the most austere assumptions about nature. In the more than a century since he derived them, they have been precisely confirmed by experiment after experiment. Yet his June 1905 paper on relativity was rejected when he submitted it as a dissertation. (He then submitted his April paper, on the size of atoms, which he thought would be less likely to startle the examiners; they accepted it only after he added one sentence to meet the length threshold.) When Einstein was awarded the 1921 Nobel Prize in Physics, it was for his work on the photoelectric effect. The Swedish Academy forbade him to make any mention of relativity in his acceptance speech. As it happened, Einstein was unable to attend the ceremony in Stockholm. He gave his Nobel lecture in Gothenburg, with King Gustav V seated in the front row. The king wanted to learn about relativity, and Einstein obliged him.

■

In 1906, the year after Einstein's annus mirabilis, Kurt Gödel was born in the city of Brno (now in the Czech Republic). Kurt was both an in-quisitive child—his parents and brother gave him the nickname *der Herr Warum*, "Mr. Why?"—and a nervous one. At the age of five, he seems to have suffered a mild anxiety neurosis. At eight, he had a terri-fying bout of rheumatic fever, which left him with the lifelong convic-tion that his heart had been fatally damaged.

Gödel entered the University of Vienna in 1924. He had intended to study physics, but he was soon seduced by the beauties of mathematics, and especially by the notion that abstractions like numbers and circles had a perfect, timeless existence independent of the human mind. This doctrine, which is called Platonism, because it descends from Plato's theory of ideas, has always been popular among mathematicians. In the philosophical world of 1920s Vienna, however, it was considered distinctly old-fashioned. Among the many intellectual movements that flourished in the city's rich café culture, one of the most prominent was the Vienna Circle, a group of thinkers united in their belief that phi-losophy must be cleansed of metaphysics and made over in the image of science. Under the influence of Ludwig Wittgenstein, their reluctant guru, the members of the Vienna Circle regarded mathematics as a game played with symbols, a more intricate version of chess. What made a proposition like "$2 + 2 = 4$" true, they held, was not that it correctly described some abstract world of numbers but that it could be derived in a logical system according to certain rules.

Gödel was introduced into the Vienna Circle by one of his professors, but he kept quiet about his Platonist views. Being both rigorous and averse to controversy, he did not like to argue his convictions unless he had an airtight way of demonstrating that they were valid. But how could one demonstrate that mathematics could not be reduced to the artifices of logic? Gödel's strategy—one of preternatural cleverness and, in the words of the philosopher Rebecca Goldstein, "heart-stopping beauty"—was to use logic against itself. Beginning with a logical system for mathematics, a system presumed to be free from contradictions, he invented an ingenious scheme that allowed the formulas in it to en-gage in a sort of doublespeak. A formula that said something about numbers could also, in this scheme, be interpreted as saying something about other formulas and how they were logically related to one another.

In fact, as Gödel showed, a numerical formula could even be made to say something about *itself*. Having painstakingly built this apparatus of mathematical self-reference, Gödel came up with an astonishing twist: he produced a formula that, while ostensibly saying something about numbers, also says, "I am not provable." At first, this looks like a paradox, recalling as it does the proverbial Cretan who announces, "All Cretans are liars." But Gödel's self-referential formula comments on its provability, not on its truthfulness. Could it be lying when it asserts, "I am not provable"? No, because if it were, that would mean it could be proved, which would make it true. So, in asserting that it cannot be proved, it has to be telling the truth. But the truth of this proposition can be seen only from outside the logical system. Inside the system, it is neither provable nor disprovable. The system, then, is incomplete, because there is at least one true proposition about numbers (the one that says "I am not provable") that cannot be proved within it. The conclusion— that no logical system can capture all the truths of mathematics—is known as the first incompleteness theorem. Gödel also proved that no logical system for mathematics could, by its own devices, be shown to be free from inconsistency, a result known as the second incompleteness theorem.

Wittgenstein once averred that "there can never be surprises in logic." But Gödel's incompleteness theorems did come as a surprise. In fact, when the fledgling logician presented them at a conference in the German city of Königsberg in 1930, almost no one was able to make any sense of them. What could it mean to say that a mathematical proposition was true if there was no possibility of proving it? The very idea seemed absurd. Even the once great logician Bertrand Russell was baffled; he seems to have been under the misapprehension that Gödel had detected an inconsistency in mathematics. "Are we to think that 2 + 2 is not 4, but 4.001?" Russell asked decades later in dismay, adding that he was "glad [he] was no longer working at mathematical logic." As the significance of Gödel's theorems began to sink in, words like "debacle," "catastrophe," and "nightmare" were bandied about. It had been an article of faith that armed with logic, mathematicians could in principle resolve any conundrum at all—that in mathematics, as it had been famously declared, there was no *ignorabimus*. Gödel's theorems seemed to have shattered this ideal of complete knowledge.

That was not the way Gödel saw it. He believed he had shown that mathematics has a robust reality that transcends any system of logic. But logic, he was convinced, is not the only route to knowledge of this reality; we also have something like an extrasensory perception of it, which he called "mathematical intuition." It is this faculty of intuition that allows us to see, for example, that the formula saying "I am not provable" must be true, even though it defies proof within the system where it lives. Some thinkers (like the physicist Roger Penrose) have taken this theme further, maintaining that Gödel's incompleteness theorems have profound implications for the nature of the human mind. Our mental powers, it is argued, must outstrip those of any computer, because a computer is just a logical system running on hardware and our minds can arrive at truths that are beyond the reach of a logical system.

Gödel was twenty-four when he proved his incompleteness theorems (a bit younger than Einstein was when he created relativity theory). At the time, much to the disapproval of his strict Lutheran parents, he was courting an older Catholic divorcée by the name of Adele, who, to top things off, was employed as a dancer in a Viennese nightclub called Der Nachtfalter (the Moth). The political situation in Austria was becoming ever more chaotic with Hitler's rise to power in Germany, although Gödel seems scarcely to have noticed. In 1936, the Vienna Circle dissolved after its founder was assassinated by a deranged student. Two years later came the Anschluss. The perilousness of the times was finally borne in upon Gödel when a band of Nazi youths roughed him up and knocked off his glasses, before retreating under the umbrella blows of Adele. He resolved to leave for Princeton, where he had been offered a position by the Institute for Advanced Study. But, the war having broken out, he judged it too risky to cross the Atlantic. So the now married couple took the long way around, traversing Russia, the Pacific, and the United States and finally arriving in Princeton in early 1940. At the institute, Gödel was given an office almost directly above Einstein's. For the rest of his life, he rarely left Princeton, which he came to find "ten times more congenial" than his once beloved Vienna.

Although Gödel was still little known in the world at large, he had a godlike status among the cognoscenti. "There it was, inconceivably, *K. Goedel*, listed just like any other name in the bright orange Princeton community phonebook," writes Rebecca Goldstein, who came to

Princeton University as a graduate student of philosophy in the early 1970s, in her intellectual biography *Incompleteness: The Proof and Paradox of Kurt Gödel* (2005). "It was like opening up the local phonebook and finding *B. Spinoza* or *I. Newton*." She goes on to recount how she "once found the philosopher Richard Rorty standing in a bit of a daze in Davidson's food market. He told me in hushed tones that he'd just seen Gödel in the frozen food aisle."

So naive and otherworldly was the great logician that Einstein felt obliged to help look after the practical aspects of his life. One much-retailed story concerns Gödel's decision after the war to become an American citizen. Gödel took the matter of citizenship with great solemnity, preparing for the exam by making a close study of the U.S. Constitution. On the appointed day, Einstein accompanied him to the courthouse in Trenton and had to intervene to quiet Gödel down when the agitated logician began explaining to the judge how the U.S. Constitution contained a loophole that would allow a dictatorship to come into existence.*

Around the same time that Gödel was studying the Constitution, he was also taking a close look at Einstein's relativity theory. The key principle of relativity is that the laws of physics should be the same for all observers. When Einstein first formulated the principle in his revolutionary 1905 paper, he restricted "all observers" to those who were moving uniformly relative to one another—that is, in a straight line and at a constant speed. But he soon realized that this restriction was arbitrary. If the laws of physics were to provide a truly objective description of nature, they ought to be valid for observers moving in any way relative to one another—spinning, accelerating, spiraling, whatever. It was thus that Einstein made the transition from his "special" theory of relativity of 1905 to his "general" theory, whose equations he worked out over the next decade and published in 1916. What made those equations so powerful was that they explained gravity, the force that governs the overall shape of the cosmos.

Decades later, Gödel, walking with Einstein, had the privilege of picking up the subtleties of relativity theory from the master himself.

---

*For more on this incident, see "Gödel Takes On the U.S. Constitution" in this volume.

Einstein had shown that the flow of time depended on motion and gravity and that the division of events into "past" and "future" was relative. Gödel took a more radical view: he believed that time, as it was intuitively understood, did not exist at all. As usual, he was not content with a mere verbal argument. Philosophers ranging from Parmenides, in ancient times, to Immanuel Kant, in the eighteenth century, and on to J.M.E. McTaggart, at the beginning of the twentieth century, had produced such arguments, inconclusively. Gödel wanted a proof that had the rigor and certainty of mathematics. And he saw just what he wanted lurking within relativity theory. He presented his argument to Einstein for his seventieth birthday, in 1949, along with an etching. (Gödel's wife had knit Einstein a sweater, but she decided not to send it.)

What Gödel found was the possibility of a hitherto unimaginable kind of universe. The equations of general relativity can be solved in a variety of ways. Each solution is, in effect, a model of how the universe might be. Einstein, who believed on philosophical grounds that the universe was eternal and unchanging, had tinkered with his equations so that they would yield such a model—a move he later called "my greatest blunder." Another physicist (a Jesuit priest, as it happens) found a solution corresponding to an expanding universe born at some moment in the finite past. Because this solution, which has come to be known as the big bang model, was consistent with what astronomers observed, it seemed to be the one that described the actual cosmos.

But Gödel came up with a third kind of solution to Einstein's equations, one in which the universe was not expanding but rotating. (The centrifugal force arising from the rotation was what kept everything from collapsing under the force of gravity.) An observer in this universe would see all the galaxies slowly spinning around him; he would know it was the universe doing the spinning and not himself, because he would feel no dizziness. What makes this rotating universe truly weird, Gödel showed, is the way its geometry mixes up space and time. By completing a sufficiently long round trip in a rocket ship, a resident of Gödel's universe could travel back to any point in his own past.

Einstein was not entirely pleased with the news that his equations permitted something as Alice in Wonderland–like as spatial paths that looped backward in time; in fact, he confessed to being "disturbed" by Gödel's universe. Other physicists marveled that time travel, previously the stuff of science fiction, was apparently consistent with the laws of physics. (Then they started worrying about what would happen if you went back to a time before you were born and killed your own grandfather.) Gödel himself drew a different moral. If time travel is possible, he submitted, then time itself is impossible. A past that can be revisited has not really passed. And the fact that the actual universe is expanding, rather than rotating, is irrelevant. Time, like God, is either necessary or nothing; if it disappears in one possible universe, it is undermined in every possible universe, including our own.

Gödel's strange cosmological gift was received by Einstein at a bleak time in his life. Einstein's quest for a unified theory of physics was proving fruitless, and his opposition to quantum theory alienated him from the mainstream of physics. Family life provided little consolation. His two marriages had been failures; a daughter born out of wedlock seems to have disappeared from history; of his two sons, one was schizophrenic, the other estranged. Einstein's circle of friends had shrunk to Gödel and a few others. One of them was Queen Elisabeth of Belgium, to whom he confided, in March 1955, that "the exaggerated esteem in which my lifework is held makes me very ill at ease. I feel compelled to think of myself as an involuntary swindler." He died a month later, at the age of seventy-six. When Gödel and another colleague went to Einstein's office at the institute to deal with his papers, they found the blackboard covered with dead-end equations.

After Einstein's death, Gödel became ever more withdrawn. He preferred to conduct all conversations by telephone, even if his interlocutor was a few feet distant. When he especially wanted to avoid someone, he would schedule a rendezvous at a precise time and place and then make sure he was somewhere far away. The honors the world wished to bestow upon him made him chary. He had shown up to collect an honorary doctorate in 1953 from Harvard, where his incompleteness theorems were hailed as the most important mathematical discovery of the previous hundred years, but he later complained of being "thrust quite undeservedly into the most highly bellicose company" of John Foster Dulles, a co-honoree. When he was awarded the National Medal of Science in 1974, he refused to go to Washington to meet Gerald Ford at the White House, despite the offer of a chauffeur for him and his wife. He had hallucinatory episodes and talked darkly of certain forces at work in the world "directly submerging the good." Fearing that there was a plot to poison him, he persistently refused to eat. Finally, looking like (in the words of a friend) "a living corpse," he was taken to the Princeton Hospital. There, two weeks later, on January 14, 1978, he succumbed to self-starvation. According to his death certificate, the cause of death was "malnutrition and inanition" brought on by "personality disturbance."

A certain futility marked the last years of both Gödel and Einstein. What might have been most futile, however, was their willed belief in the unreality of time. The temptation was understandable. If time is merely in our minds, perhaps we can hope to escape it into a timeless eternity. Then we could say, like William Blake, "I see the Past, Present, and Future existing all at once / Before me." In Gödel's case, it might have been his childhood terror of a fatally damaged heart that attracted him to the idea of a timeless universe. Toward the end of his life, he told one confidant that he had long awaited an epiphany that would enable him to see the world in a new light but that it never came.

Einstein, too, was unable to make a clean break with time. "To those of us who believe in physics," he wrote to the widow of a friend who had recently died, "this separation between past, present, and future is only an illusion, if a stubborn one." When his own turn came, a couple of weeks later, he said, "It is time to go."

2

Time—the Grand Illusion?

Isaac Newton had a peculiar notion of time. He saw it as a sort of cosmic grandfather clock, one that hovered over the rest of nature in blithe autonomy. And he believed that time advanced at a smooth and constant rate from past to future. "Absolute, true, mathematical time, of itself, and from its own nature, flows equably without relation to anything external," Newton declared at the beginning of his *Principia*. To those caught up in the temporal flux of daily life, this seems like arrant nonsense. Time does not strike us as transcendent and mathematical; rather, it is something intimate and subjective. Nor does it proceed at a stately and unvarying pace. We know that time has different tempos. In the run-up to New Year's Eve, for instance, time positively flies. Then, in January and February, it slows to a miserable crawl. Moreover, time moves faster for some of us than for others. Old people are being rushed forward into the future at a cruelly rapid clip. When you're an adult, as Fran Lebowitz once observed, Christmas seems to come every five minutes. For little children, however, time goes quite slowly. Owing to the endless novelty of a child's experience, a single summer can stretch out into an eternity. It has been estimated that by the age of eight, one has subjectively lived two-thirds of one's life.

Researchers have tried to measure the subjective flow of time by asking people of different ages to estimate when a certain amount of time has gone by. People in their early twenties tend to be quite accurate in judging when three minutes had elapsed, typically being off by no more than three seconds. Those in their sixties, by contrast, over-shot the mark by forty seconds; in other words, what was actually three minutes and forty seconds seemed like only three minutes to them. Seniors are internally slow tickers, so for them actual clocks seem to tick too fast. This can have its advantages: at a John Cage concert, it is the old people who are relieved that the composition 4′33″ is over so soon.

The river of time may have its rapids and its calmer stretches, but one thing would seem to be certain: it carries all of us, willy-nilly, in its flow. Irresistibly, irreversibly, we are being borne toward our deaths at the stark rate of one second per second. As the past slips out of existence behind us, the future, once unknown and mysterious, assumes its banal reality before us as it yields to the ever-hurrying *now*.

But this sense of flow is a monstrous illusion—so says contemporary physics. And Newton was as much a victim of this illusion as the rest of us are.

It was Albert Einstein who initiated the revolution in our understanding of time. In 1905, Einstein showed that time, as it had been understood by physicist and plain man alike, was a fiction. Einstein proved that whether an observer deems two events at different locations to be happening "at the same time" depends on his state of motion. Suppose, for example, that Jones is walking uptown on Fifth Avenue and Smith is walking downtown. Their relative motion results in a discrepancy of several days in what they would judge to be happening "now" in the Andromeda galaxy at the moment they pass each other on the sidewalk. For Smith, the space fleet launched to destroy life on earth is already on its way; for Jones, the Andromedan council of tyrants has not even decided whether to send the fleet.

What Einstein had shown was that there is no universal "now." Whether two events are simultaneous is relative to the observer. And once simultaneity goes by the board, the very division of moments into "past," "present," and "future" becomes meaningless. Events judged to be in the past by one observer may still lie in the future of another; therefore, past and present must be equally definite, equally "real." In place of the fleeting present, we are left with a vast frozen timescape—a four-dimensional "block universe." Over here, you are being born; over there, you are celebrating the turn of the millennium; and over yonder, you've been dead for a while. Nothing is "flowing" from one event to another. As the mathematician Hermann Weyl memorably put it, "The objective world simply is; it does not happen."

Einstein, through his theory of relativity, furnished a scientific justification for a philosophical view of time that goes back to Spinoza, to Saint Augustine, even to Parmenides—one that has been dubbed eternalism. Time, according to this view, belongs to the realm of appearance, not reality. The only objective way to see the universe is as God sees it: *sub specie aeternitatis*.

Over the decades since Einstein's death, physics has subjected our everyday notion of time to still more radical dislocations. The frozen timescape of relativity theory has been revealed to have gaping holes in it: black holes. That is because time is "warped" by gravity. The stronger the gravitational field, the slower the clock hands creep. If you live in a ground-floor apartment, you age a trifle less rapidly than your neighbor in the penthouse. The effect would be a lot more noticeable if you got sucked into a black hole, where the gravitational warpage of time is infinite. Quite literally, black holes are gateways to the end of time: to Nowhen.

If time behaves dodgily around black holes, it may vanish altogether at the tiniest of scales, where the fabric of space-time dissolves into a "quantum foam" in which events have no determinate temporal order. Temporal matters are even stranger if we look back at the big bang, the cataclysmic event that ushered our universe into existence—and not just the universe but also its space-time container. We all want to ask, what the heck was going on just before the big bang? But that question is nonsensical. So, at any rate, we are told by Stephen Hawking in *A Brief History of Time*. Invoking what he calls "imaginary time"—a notion that has been known to puzzle even his fellow physicists—Hawking says that asking what came before the big bang is as silly as asking what's north of the North Pole. The answer, of course, is nothing.

Does time have a future? Yes, but how much of a future depends on what the ultimate fate of the cosmos turns out to be. The possibilities come down to Robert Frost's choice: fire or ice? Ever since its birth in the big bang, some 13.82 billion years ago, the universe has been expanding. If this expansion continues forever, the universe will end in ice, at least metaphorically speaking. The stars will burn out; black holes will evaporate; atoms and their subatomic constituents will decay. In the deep future, the remaining particles (mainly photons and neutrinos) will spread out into the void, becoming so distant from one another that they will cease to interact. Space will become empty except for the merest hint of "vacuum energy." Yet in this future wasteland of near nothingness, time will go on; random events will continue to occur; things will "fluctuate" into existence, thanks to the magic of quantum uncertainty, only to disappear again into the void. Most of these future ephemera will be single particles, like electrons and protons. But every once in a while—a very great while—more complicated structures will spontaneously wink into being: say, a human brain. Indeed, in the fullness of time, quantum physics could allow for an infinite number of such disembodied brains, stocked with (false) memories, that will appear and disappear. In the scientific literature, these sad and evanescent entities are called Boltzmann brains (after Ludwig Boltzmann, one of the pioneers of modern thermodynamics). One such deep-future Boltzmann brain would be identical to your own brain as it is constituted at this very moment. Thus, in some inconceivably distant epoch, your current state of consciousness would be re-created out of the void, only to be extinguished an instant later—not, perhaps, the kind of resurrection you were hoping for.

All that could be true (says current physics) provided the universe continues to expand eternally, growing ever emptier and darker and colder: a scenario that might be called the big chill. But there is another possible cosmic fate. By and by, at some point in the far future, the expansion that the universe is currently undergoing might be arrested—maybe by gravity, maybe by some force that is currently unknown. Then all the hundreds of billions of galaxies will begin to collapse back on themselves, eventually coming together in a fiery all-annihilating implosion: the big crunch. Just as the big bang brought time into existence, the big crunch would bring it to an end. Or would it? Some cosmic optimists have argued that in the final moments before such a big crunch an infinite amount of energy would be released. This energy, the optimists say, could be harnessed by our deep-future descendants to power an infinite amount of computation, giving rise to an infinite number of thoughts. Because these thoughts would unfold at a faster and faster pace, subjective time would seem to go on forever, even though objective time was about to come to an end. The split second before the big crunch would thus be like a child's endless summer: a virtual eternity.

Virtual eternity, gateways to Nowhen, the unreality of time . . . Do any of these lotus-eater ideas really hit us where we live, in the life-world? Probably not. Like Einstein himself, we are stubbornly in thrall to our temporal illusions. We cannot help feeling ourselves to be slaves to one part of the timescape (the past) and hostages to another part (the future). Nor can we help feeling that we are quite literally running out of time. Arthur Eddington, one of the first physicists to grasp Einstein's relativity theory, declared that our intuitive sense of time's passage is so powerful that it must correspond to something in the objective world. If science cannot get a purchase on it, one might say, well, so much the worse for science!

What science *can* tell us something about is the psychology of time's passage. Our conscious now—what William James dubbed the "specious present"—is actually an interval of about three seconds. That is the span over which our brains knit up arriving sense data into a unified experience. It is also pretty clear that the nature of memory has something to do with the feeling that we are moving in time. The past and the future might be equally real, but—for reasons traceable, oddly enough, to the second law of thermodynamics—we cannot "remember" events in the future, only ones in the past. Memories accumulate in one temporal direction and not in the other. This seems to explain the psychological arrow of time. It does not, unfortunately, explain why that arrow seems to fly.

If all of this leaves you utterly bewildered about time, you are in eminent company. John Archibald Wheeler, one of the great physicists of the twentieth century, took to quoting this in a scientific paper: "Time is nature's way to keep everything from happening all at once." In a footnote, Wheeler writes that he discovered this quotation among graffiti in the men's room at the Old Pecan Street Café in Austin, Texas. That such a distinguished thinker would resort to quoting from a men's room wall isn't surprising if you consider the contemporary free-for-all among physicists and philosophers and philosophers of physics over the nature of time. Some maintain that time is a basic ingredient of the universe; others say, no, it emerges from deeper features of physical reality. Some insist that time has a built-in direction; others deny this.

(Stephen Hawking once claimed that time could eventually reverse it-self and run backward, only to realize later that there had been a mistake in his calculations.) Most physicists and philosophers today agree with Einstein that time's passage is an illusion; they are eternal-ists. But a minority—who call themselves presentists—think that *now* is a special moment that really advances, like a little light moving along the line of history; this would still be true, they believe, even if there were no observers like us in the universe.

If there is one proposition about time that all scientifically inclined thinkers can agree on, it might be one attributed to the nonscientist Hector Berlioz, who is said to have quipped, "Time is a great teacher, but unfortunately it kills all its pupils."

PART II

Numbers in the Brain,
in Platonic Heaven,
and in Society

3

Numbers Guy:
The Neuroscience of Math

One morning in September 1989, a former sales representative in his mid-forties entered an examination room with Stanislas Dehaene, a young neuroscientist based in Paris. Three years earlier, the man, whom researchers came to refer to as Mr. N., had sustained a brain hemorrhage that left him with an enormous lesion in the rear half of his left hemisphere. He suffered from severe handicaps: his right arm was in a sling; he couldn't read; and his speech was painfully slow. He had once been married, with two daughters, but was now incapable of leading an independent life and lived with his elderly parents. Dehaene had been invited to see him because his impairments included severe acalculia, a general term for any one of several deficits in number processing. When asked to add 2 and 2, he answered "Three." He could still count and recite a sequence like 2, 4, 6, 8, but he was incapable of counting downward from 9, differentiating odd and even numbers, or recognizing the numeral 5 when it was flashed in front of him.

To Dehaene, these impairments were less interesting than the fragmentary capabilities Mr. N. had managed to retain. When he was shown the numeral 5 for a few seconds, he knew it was a numeral rather than a letter and, by counting up from 1 until he got to the right integer, he eventually identified it as a 5. He did the same thing when asked the age of his seven-year-old daughter. In the 1997 book *The Number Sense*, Dehaene wrote, "He appears to know right from the start what quantities he wishes to express, but reciting the number series seems to be his only means of retrieving the corresponding word."

Dehaene also noticed that although Mr. N. could no longer read, he sometimes had an approximate sense of words that were flashed in front of him; when he was shown the word "ham," he said, "It's some kind of meat." Dehaene decided to see if Mr. N. still had a similar sense of number. He showed him the numerals 7 and 8. Mr. N. was able to answer quickly that 8 was the larger number—far more quickly than if he had had to identify them by counting up to the right quantities. He could also judge whether various numbers were bigger or smaller than 55, slipping up only when they were very close to 55. Dehaene dubbed Mr. N. "the Approximate Man." The Approximate Man lived in a world where a year comprised "about 350 days" and an hour "about fifty minutes," where there were five seasons, and where a dozen eggs amounted to "six or ten." Dehaene asked him to add 2 and 2 several times and received answers ranging from 3 to 5. But, he noted, "he never offers a result as absurd as 9."

In cognitive science, incidents of brain damage are nature's experiments. If a lesion knocks out one ability but leaves another intact, it is evidence that they are wired into different neural circuits. In this instance, Dehaene theorized that our ability to learn sophisticated mathematical procedures resided in an entirely different part of the brain from a rougher quantitative sense. Over the decades, evidence concerning cognitive deficits in brain-damaged patients has accumulated, and researchers have concluded that we have a sense of number that is independent of language, memory, and reasoning in general. Within neuroscience, numerical cognition has emerged as a vibrant field, and Dehaene has become one of its foremost researchers. His work is "completely pioneering," Susan Carey, a psychology professor at Harvard who has studied numerical cognition, told me. "If you want to make sure the math that children are learning is meaningful, you have to know something about how the brain represents number at the kind of level that Stan is trying to understand."

Dehaene has spent most of his career plotting the contours of our number sense and puzzling over which aspects of our mathematical ability are innate and which are learned, and how the two systems overlap and affect each other. He has approached the problem from every imaginable angle. Working with colleagues both in France and in the United States, he has carried out experiments that probe the way numbers are coded in our minds. He has studied the numerical abilities of animals, of Amazon tribespeople, of top French mathematics students. He has used brain-scanning technology to investigate precisely where in the folds and crevices of the cerebral cortex our numerical faculties are nestled. And he has weighed the extent to which some languages make numbers more difficult than others.

Dehaene's work raises crucial issues about the way mathematics is taught. In his view, we are all born with an evolutionarily ancient mathematical instinct. To become numerate, children must capitalize on this instinct, but they must also unlearn certain tendencies that were helpful to our primate ancestors but that clash with skills needed today. And some societies are evidently better than others at getting kids to do this. In both France and the United States, mathematics education is often felt to be in a state of crisis. The math skills of American children fare poorly in comparison with those of their peers in countries like Singapore, South Korea, and Japan. Fixing this state of affairs means grappling with the question that has taken up much of Dehaene's career: What is it about the brain that makes numbers sometimes so easy and sometimes so hard?

Dehaene's own gifts as a mathematician are considerable. Born in 1965, he grew up in Roubaix, a medium-sized industrial city near France's border with Belgium. (His surname is Flemish.) His father, a pediatrician, was among the first to study fetal alcohol syndrome. As a teenager, Dehaene developed what he calls a "passion" for mathematics, and he attended the École Normale Supérieure in Paris, the training ground for France's scholarly elite. Dehaene's own interests tended toward computer modeling and artificial intelligence. He was drawn to brain science after reading, at the age of eighteen, the 1983 book *Neuronal Man*, by Jean-Pierre Changeux, France's most distinguished neurobiologist. Changeux's approach to the brain held out the tantalizing possibility of reconciling psychology with neuroscience. Dehaene met Changeux and began to work with him on abstract models of thinking and memory. He also linked up with the cognitive scientist Jacques Mehler. It was in Mehler's lab that he met his future wife, Ghislaine Lambertz, a researcher in infant cognitive psychology.

By "pure luck," Dehaene recalls, Mehler happened to be doing research on how numbers are understood. This led to Dehaene's first encounter with what he came to characterize as "the number sense." Dehaene's work centered on an apparently simple question: How do we know whether numbers are bigger or smaller than each other? If you are asked to choose which of a pair of Arabic numerals—4 and 7, say—stands for the bigger number, you respond "7" in a split second, and one might think that any two digits could be compared in the same very brief period of time. Yet in Dehaene's experiments, while subjects answered quickly and accurately when the digits were far apart, like 2 and 9, they slowed down when the digits were closer together, like 5 and 6. Performance also got worse as the digits grew larger: 2 and 3 were much easier to compare than 7 and 8. When Dehaene tested some of the best mathematics students at the École Normale, the students were amazed to find themselves slowing down and making errors when asked whether 8 or 9 was the larger number.

Dehaene conjectured that when we see numerals or hear number words, our brains automatically map them onto a number line that grows increasingly fuzzy above 3 or 4. He found that no amount of training can change this. "It is a basic structural property of how our brains represent number, not just a lack of facility," he told me.

In 1987, while Dehaene was still a student in Paris, the American cognitive psychologist Michael Posner and colleagues at Washington University in St. Louis published a pioneering paper in the journal *Nature*. Using a scanning technique that can track the flow of blood in the brain, Posner's team had detailed how different areas became active in language processing. Their research was a revelation for Dehaene. "I remember very well sitting and reading this paper and then debating it with Jacques Mehler, my Ph.D. adviser," he told me. Mehler, whose focus was on determining the abstract organization of cognitive functions, didn't see the point of trying to locate precisely where in the brain things happened, but Dehaene wanted to "bridge the gap," as he put it, between psychology and neurobiology, to find out exactly how the functions of the mind—thought, perception, feeling, will—are realized in the gelatinous three-pound lump of matter in our skulls. Now, thanks to new technologies, it was finally possible to create pictures, however crude, of the brain in the act of thinking. So, after receiving his doctorate, he spent two years studying brain scanning with Posner, who was by then at the University of Oregon, in Eugene. "It was very strange to find that some of the most exciting results of the budding cognitive-neuroscience field were coming out of this small place—the only place where I ever saw sixty-year-old hippies sitting around in tie-dyed shirts!" he said.

Dehaene is a compact, attractive, and genial man; he dresses casually, wears fashionable glasses, and has a glabrous dome of a head, which he protects from the elements with a *chapeau de cowboy*. When I first visited him back in 2008, he had just moved into a new laboratory, known as NeuroSpin, on the campus of a national center for nuclear-energy research, a dozen or so miles southwest of Paris. The building is a modernist composition in glass and metal filled with the ambient hums and whirs and whooshes of brain-scanning equipment, much of which was still being assembled at the time. A series of arches ran along one wall in the form of a giant sine wave; behind each was a concrete vault built to house a liquid-helium-cooled superconducting electromagnet. (In brain imaging, the more powerful the magnetic field, the sharper the picture.) The new brain scanners were expected to show the human cerebral anatomy at a level of detail never before seen, perhaps revealing subtle anomalies in the brains of people with dyslexia and with dyscalculia, a crippling deficiency in dealing with numbers that, researchers suspect, may be as widespread as dyslexia.

One of the scanners was already up and running. "You don't wear a pacemaker or anything, do you?" Dehaene asked me as we entered a room where two researchers were fiddling with controls. Although the scanner was built to accommodate humans, inside, I could see from the monitor, was a brown rat. Researchers were looking at how its brain reacted to various odors, which were puffed in every so often. Then Dehaene led me upstairs to a spacious gallery where the brain scientists working at NeuroSpin are expected to congregate and share ideas. At the moment, it was empty. "We're hoping for a coffee machine," he said.

Dehaene has established himself as an international scanning virtuoso. On returning to France after his time with Posner, he pressed on with the use of imaging technologies to study how the mind processes numbers. The existence of an evolved number ability had long been hypothesized, based on research with animals and infants, and evidence from brain-damaged patients gave clues to where in the brain it might be found. Dehaene set about localizing this facility more precisely and describing its architecture. "In one experiment I particularly liked," he recalled, "we tried to map the whole parietal lobe in a half hour, by having the subject perform functions like moving the eyes and hands, pointing with fingers, grasping an object, engaging in various language tasks, and, of course, making small calculations, like thirteen minus four. We found there was a beautiful geometrical organization to the areas that were activated. The eye movements were at the back, the hand movements were in the middle, grasping was in the front, and so on. And right in the middle, we were able to confirm, was an area that cared about number."

The number area lies deep within a fold in the parietal lobe called the intraparietal sulcus (just behind the crown of the head). But it isn't easy to tell what the neurons there are actually doing. Brain imaging, for all the sophistication of its technology, yields a fairly crude picture of what's going on inside the skull, and the same spot in the brain might light up for two tasks, even though different neurons are involved. "Some people believe that psychology is just being replaced by brain imaging, but I don't think that's the case at all," Dehaene said. "We need psychology to refine our idea of what the imagery is going to show us. That's why we do behavioral experiments, see patients. It's the confrontation of all these different methods that creates knowledge."

Dehaene has been able to bring together the experimental and the theoretical sides of his quest, and on at least one occasion he has even theorized the existence of a neurological feature whose presence was later confirmed by other researchers. In the early 1990s, working with Jean-Pierre Changeux, he set out to create a computer model to simulate the way humans and some animals estimate at a glance the number of objects in their environment. In the case of very small numbers, this estimate can be made with almost perfect accuracy, an ability known as subitizing (from the Latin word *subitus*, meaning "sudden"). Some psychologists think that subitizing is merely rapid, unconscious counting, but others, Dehaene included, believe that our minds perceive up to three or four objects all at once, without having to mentally "spotlight" them one by one.

Getting the computer model to subitize the way humans and animals did was possible, Dehaene found, only if he built in "number neurons" tuned to fire with maximum intensity in response to a specific number of objects. His model had, for example, a special four neuron that got particularly excited when the computer was presented with four objects. The model's number neurons were pure theory, but almost a decade later two teams of researchers discovered what seemed to be the real item, in the brains of macaque monkeys that had been trained to do number tasks. The number neurons fired precisely the way Dehaene's model predicted—a stunning vindication of theoretical psychology. "Basically, we can derive the behavioral properties of these neurons from first principles," he told me. "Psychology has become a little more like physics."

But the brain is the product of evolution—a messy, random process—and though the number sense may be lodged in a particular bit of the cerebral cortex, its circuitry seems to be intermingled with the wiring for other mental functions. A few years ago, while analyzing an experiment on number comparisons, Dehaene noticed that subjects performed better with large numbers if they held the response key in their right hand but did better with small numbers if they held the response key in their left hand. Strangely, if the subjects were made to cross their hands, the effect was reversed. The actual hand used to make the response was, it seemed, irrelevant; it was space itself that the subjects unconsciously associated with larger or smaller numbers. Dehaene hypothesizes that the neural circuitry for number and the circuitry for location overlap. He even suspects that this may be why travelers get disoriented entering Terminal 2 of Paris's Charles de Gaulle Airport, where small-numbered gates are on the right and large-numbered gates are on the left. "It's become a whole industry now to see how we associate number to space and space to number," Dehaene said. "And we're finding the association goes very, very deep in the brain."

Later, I accompanied Dehaene to the ornate setting of the Institut de France, across the Seine from the Louvre. There he accepted a prize of a quarter of a million euros from Liliane Bettencourt, whose father created the cosmetics group L'Oréal. In a salon hung with baroque tapestries, Dehaene described his research to a small audience that included a former prime minister of France. New techniques of neuroimaging, he explained, promise to reveal how a thought process like calculation unfolds in the brain. This isn't just a matter of pure knowledge, he added. Because the brain's architecture determines the sorts of abilities that come naturally to us, a detailed understanding of that architecture should lead to better ways of teaching children mathematics and may help close the educational gap that separates children in the West from those in several Asian countries.

The fundamental problem with learning mathematics is that while the number sense may be genetic, exact calculation requires cultural tools—symbols and algorithms—that have been around for only a few thousand years and must therefore be absorbed by areas of the brain that evolved for other purposes. The process is made easier when what we are learning harmonizes with built-in circuitry. If we can't change the architecture of our brains, we can at least adapt our teaching methods to the constraints it imposes.

For nearly three decades, American educators have pushed "reform math," in which children are encouraged to explore their own ways of solving problems. Before reform math, there was the "new math," now widely thought to have been an educational disaster. (In France, it was called *les maths modernes* and is similarly despised.) The new math was grounded in the theories of the influential Swiss psychologist Jean Piaget, who believed that children are born without any sense of number and only gradually build up the concept in a series of developmental stages. Piaget thought that children, until the age of four or five, cannot grasp the simple principle that moving objects around does not affect how many of them there are, and that there was therefore no point in trying to teach them arithmetic before the age of six or seven.

Piaget's view had become standard by the 1950s, but psychologists have since come to believe that he underrated the arithmetic competence of small children. Six-month-old babies, exposed simultaneously to images of common objects and sequences of drumbeats, consistently gaze longer at the collection of objects that matches the number of drumbeats. By now, it is generally agreed that infants come equipped with a rudimentary ability to perceive and represent number. (The same appears to be true for many kinds of animals, including salamanders, pigeons, raccoons, dolphins, parrots, and monkeys.) And if evolution has equipped us with one way of representing number, embodied in the primitive number sense, culture furnishes two more: numerals and number words. These three modes of thinking about number, Dehaene believes, correspond to distinct areas of the brain. The number sense is lodged in the parietal lobe, the part of the brain that relates to space and location; numerals are dealt with by the visual areas; and number words are processed by the language areas.

Nowhere in all this elaborate brain circuitry, alas, is there the equivalent of the chip found in a five-dollar calculator. This deficiency can make learning that terrible quartet—"Ambition, Distraction, Uglification, and Derision," as Lewis Carroll burlesqued them—a chore. It's not so bad at first. Our number sense endows us with a crude feel for addition, so that, even before schooling, children can find simple recipes for adding numbers. If asked to compute $2 + 4$, for example, a child might start with the first number and then count upward by the second number: "two, three is one, four is two, five is three, six is four, six." But multiplication is another matter. It is an "unnatural practice," Dehaene is fond of saying, and the reason is that our brains are wired the wrong way. Neither intuition nor counting is of much use, and multiplication facts must be stored in the brain verbally, as strings of words. The list of arithmetical facts to be memorized may be short, but it is fiendishly tricky: the same numbers occur over and over, in different orders, with partial overlaps and irrelevant rhymes. (Bilinguals, it has been found, revert to the language they used in school when doing multiplication.) The human memory, unlike that of a computer, has evolved to be associative, which makes it ill-suited to arithmetic, where bits of knowledge must be kept from interfering with one another: if you're trying to retrieve the result of multiplying 7×6, the reflex activation of $7 + 6$ and 7×5 can be disastrous. So multiplication is a double terror: not only is it remote from our intuitive sense of number; it has to be internalized in a form that clashes with the evolved organization of our memory. The result is that when adults multiply single-digit numbers they make mistakes 10 to 15 percent of the time. For the hardest problems, like 7×8, the error rate can exceed 25 percent.

Our built-in ineptness when it comes to more complex mathematical processes has led Dehaene to question why we insist on drilling procedures like long division into our children at all. There is, after all, an alternative: the electronic calculator. "Give a calculator to a five-year-old, and you will teach him how to make friends with numbers instead of despising them," he has written. By removing the need to spend hundreds of hours memorizing boring procedures, he says, calculators can free children to concentrate on the meaning of these procedures, which is neglected under the educational status quo.

This attitude might make Dehaene sound like a natural ally of educators who advocate reform math and a natural foe of parents who want their children's math teachers to go "back to basics." But when I asked him about reform math, he wasn't especially sympathetic. "The idea that all children are different and that they need to discover things their own way—I don't buy it at all," he said. "I believe there is one brain organization. We see it in babies; we see it in adults. Basically, with a few variations, we're all traveling on the same road." He admires the mathematics curricula of Asian countries like China and Japan, which provide children with a highly structured experience, anticipating the kinds of responses they make at each stage and presenting them with challenges designed to minimize the number of errors. "That's what we're trying to get back to in France," he said. Working with his colleague Anna Wilson, Dehaene has developed a computer game called *The Number Race* to help dyscalculic children. The software is adaptive, detecting the number tasks where the child is shaky and adjusting the level of difficulty to maintain an encouraging success rate of 75 percent.

Despite our shared brain organization, cultural differences in how we handle numbers persist, and they are not confined to the classroom. Evolution might have endowed us with an approximate number line, but it takes a system of symbols to make numbers precise—to "crystallize" them, in Dehaene's metaphor. The Mundurukú, an Amazon tribe that Dehaene and colleagues, notably the linguist Pierre Pica, have studied recently, have words for numbers only up to five. (Their word for five literally means "one hand.") Even these words seem to be merely approximate labels for them: a Mundurukú who is shown three objects will sometimes say there are three, sometimes four. Nevertheless, the Mundurukú have a good numerical intuition. "They know, for example, that fifty plus thirty is going to be larger than sixty,"

Dehaene said. "Of course, they do not know this verbally and have no way of talking about it. But when we showed them the relevant sets and transformations, they immediately got it."

The Mundurukú, it seems, have developed few cultural tools to augment the inborn number sense. Interestingly, the very symbols with which we write down the counting numbers bear the trace of a similar stage. The first three Roman numerals, I, II, and III, were formed by using the symbol for one as many times as necessary; the symbol for four, IV, is not so transparent. The same principle applies to Chinese numerals: the first three consist of one, two, and three horizontal bars, but the fourth takes a different form. Even Arabic numerals follow this logic: 1 is a single vertical bar; 2 and 3 began as two and three horizontal bars tied together for ease of writing. ("That's a beautiful little fact, but I don't think it's coded in our brains any longer," Dehaene observed.)

Today, Arabic numerals are in use pretty much around the world, while the words with which we name numbers naturally differ from language to language. And, as Dehaene and others have noted, these differences are far from trivial. English is cumbersome. There are special words for the numbers from 11 to 19 and for the decades from 20 to 90. This makes counting a challenge for English-speaking children, who are prone to such errors as "twenty-eight, twenty-nine, twenty-ten, twenty-eleven." French is just as bad, with vestigial base-twenty monstrosities, like *quatre-vingt-dix-neuf* (four twenty ten nine) for 99. Chinese, by contrast, is simplicity itself; its number syntax perfectly mirrors the base-ten form of Arabic numerals, with a minimum of terms. Consequently, the average Chinese four-year-old can count up to forty, whereas American children of the same age struggle to get to fifteen. And the advantages extend to adults. Because Chinese number words are so brief—they take less than a quarter of a second to say, on average, compared with a third of a second for English—the average Chinese speaker has a memory span of nine digits, versus seven digits for English speakers. (Speakers of the marvelously efficient Cantonese dialect, common in Hong Kong, can juggle ten digits in active memory.)

In 2005, Dehaene was elected to the chair in experimental cognitive psychology at the Collège de France, a highly prestigious institution founded by Francis I in 1530. The faculty consists of just fifty-two scholars, and Dehaene became the youngest member. In his inaugural lecture, Dehaene marveled at the fact that mathematics is simultaneously a product of the human mind and a powerful instrument for discovering the laws by which the human mind operates. He spoke of the confrontation between new technologies like brain imaging and ancient philosophical questions concerning number, space, and time. And he pronounced himself lucky to be living in an era when advances in psychology and neuroimaging are combining to "render visible" the hitherto invisible realm of thought.

For Dehaene, numerical thought is only the beginning of this quest. He has spent the latter part of his career pondering how the philosophical problem of consciousness might be approached by the methods of empirical science. Experiments involving subliminal "number priming" show that much of what our mind does with numbers is unconscious, a finding that has led Dehaene to wonder why some mental activity crosses the threshold of awareness and some doesn't. Collaborating with a couple of colleagues, Dehaene has explored the neural basis of what is known as the "global workspace" theory of consciousness, which has elicited keen interest among philosophers. In his version of the theory, information becomes conscious when certain "workspace" neurons broadcast it to many areas of the brain at once, making it simultaneously available for, say, language, memory, perceptual categorization, action planning, and so on. In other words, consciousness is "cerebral celebrity," as the philosopher Daniel Dennett has described it, or "fame in the brain."

In his office at NeuroSpin, Dehaene described to me how certain extremely long workspace neurons might link far-flung areas of the human brain together into a single pulsating circuit of consciousness. To show me where these areas were, he reached into a closet and pulled out an irregularly shaped baby-blue plaster object, about the size of a softball. "This is my brain!" he announced with evident pleasure. The model that he was holding had been fabricated, he explained, by a rapid-prototyping machine (a sort of three-dimensional printer) from computer data obtained from one of the many MRI scans that he has undergone. He pointed to the little furrow where the number sense was supposed to be situated and observed that his had a somewhat uncommon shape. Curiously, the computer software had identified

Dehaene's brain as an "outlier," so dissimilar are its activation patterns from the human norm. Cradling the pastel-colored lump in his hands, a model of his mind devised by his own mental efforts, Dehaene paused for a moment. Then he smiled and said, "So, I kind of like my brain."

4

The Riemann Zeta Conjecture
and the Laughter of the Primes

What will civilization be like a million years from now? Most of the things we're familiar with today will have disappeared. But some things will survive. And we can be pretty confident that among them will be numbers and laughter. That is good, because, in their very different ways, numbers and laughter make life worth living. So it is interesting to ponder their status in the Year Million. But first let me tell you why I am so sure they will still be around, when almost everything else we know today will either have disappeared or have evolved into something quite unrecognizable.

In general, things that have been around for a long time will likely be around for a lot longer. Conversely, things of recent origin likely won't be. Both of these conclusions flow from the "Copernican principle," which says, in essence, *you're not special*. If there's nothing special about our perspective, we're unlikely to be observing any given thing at the very beginning or the very end of its existence. Let's say you go to see a Broadway play. No one can be sure exactly how long the play will run. It could be anything from a few nights to many years. But you do know that of all the people who will see it, 95 percent of them will be among neither the first 2.5 percent nor the last 2.5 percent to do so. Therefore, if you're not "special"—that is, if you're just a random member of the play's total audience—you can be 95 percent sure that you don't fall into either of these two tails. What this means is that if the play has already had *n* performances at the point in its run when you happen to see it, you can be 95 percent sure that it has no more than $39 \times n$ performances to go and no fewer than $n \div 39$. (This is a matter of elementary arithmetic: the upper limit keeps you out of the first 2.5 percent of the total audience, and the lower limit keeps you out of the last 2.5 percent.) With nothing more than the Copernican principle and a grade school calculation, you can come up with a 95 percent confidence interval for the longevity of something like a Broadway play. That is fairly amazing.

It was J. Richard Gott III, an astrophysicist at Princeton University, who pioneered this sort of reasoning. In a 1993 paper published in *Nature*, "Implications of the Copernican Principle for Our Future Prospects," Gott calculated the expected longevity of our species. Humans have already been around for about 200,000 years. So, if there is nothing special about the moment at which we observe our species, we can be 95 percent sure that it will continue to be around for at least 5,100 years ($1/39 \times 200,000$) but that it will disappear within 7.8 million years ($39 \times 200,000$). This, Gott noted, gives *Homo sapiens* an expected total longevity comparable to other hominid species (*Homo erectus*, our ancestor species, lasted 1.6 million years) and to mammal species in general (whose average span is 2 million years). It also gives us a decent shot at being around in the Year Million—although that probability is not so great as we might naively hope (see "Doom Soon," p. 259).

But what else will be around then? Consider something of recent origin, like the Internet. The Internet has existed for about a third of a century (as I learned by going on the Internet and looking at *Wikipedia*). That means, by Copernican reasoning, we can be 95 percent certain that it will continue to be around for another ten months but that it will disappear within thirteen hundred years. So, in the Year Million, there will almost certainly be nothing recognizable as the Internet. (This is, perhaps, not a terribly surprising conclusion.) Ditto for baseball, which has been around for a little more than two centuries now. Ditto for what we call industrial technology, which, having come into existence a few hundred years ago, is likely to be superseded by something strange and new in the next ten thousand years. Nor, by the same Copernican reasoning, is organized religion a good bet to survive to anything like the Year Million.

To find something that will pretty certainly still be around in the Year Million, we are obliged, paradoxically enough, to go back much further in our natural history. Again, this is because (as Gott puts it) "things that have been around for a long time tend to stay around for a long time." And if we could cast a look back several million years into the past, we would see, among other things, laughter and numbers. How do we know this? Because we share both laughter and a sense of number with other species today, and therefore shared them with common ancestors that existed millions of years ago.

Take laughter. Chimpanzees laugh. Charles Darwin notes in *The Expression of the Emotions in Man and Animals* (1872), "If a young chimpanzee be tickled—the armpits are particularly sensitive to tickling, as in the case of our children—a more decided chuckling or laughing sound is uttered; though the laughter is sometimes noiseless." Actually, what primatologists call chimp laughter is closer to a breathy pant. It is evoked not only by tickling but also by rough-and-tumble play, games of chasing, and mock attacks—just as with children prior to the emergence of verbal joking at age five or six. But does primate humor ever rise above sheer physicality? The researcher Roger Fouts reported that Washoe, a chimp who was taught sign language, once urinated on him while riding on his shoulders, signing "funny" and snorting but not laughing.

The human and chimpanzee lineages split off from each other between five and seven million years ago. On the reasonable assumption that chimp and human laughter are "homologous," rather than traits that evolved independently, that means that laughter must be at least five to seven million years old. (It is probably much older: orangutans also laugh, and their lineage diverged from ours about fourteen million years ago.) So, by the Copernican principle, laughter is quite likely to be around in the Year Million.

Now take numbers. Chimps can also do elementary arithmetic, and they have even been trained to use symbols like numerals to reason about quantity. And a sense of number is not confined to primates. Researchers have found that animals as diverse as salamanders, dolphins, and raccoons have the ability to perceive and represent number. A couple of decades ago, researchers at MIT found that macaque monkeys had specialized "number neurons" in the part of their brain that corresponds to the location of the human number module. Evidently, the number sense has an even longer evolutionary history than that of laughter. So again, by the Copernican principle, we can be quite certain that numbers will be around in the Year Million.

Of the cultural wonders of our world, numbers and laughter are two of the oldest. As such, they are likely to survive the longest, quite probably beyond the Year Million. One might make an analogy to the Seven Wonders of the Ancient World. When this list was drawn up (the earliest extant version dates to about 140 B.C.E.), the oldest wonder on it by far was the pyramid of Giza, which went back to about 2500 B.C.E. The other six wonders—the Hanging Gardens of Babylon, the temple of Artemis at Ephesus, the statue of Zeus at Olympia, the mausoleum at Halicarnassus, the Colossus of Rhodes, and the lighthouse at Alexandria—were almost two millennia newer. And which of the Seven Wonders still survives today? The pyramid of Giza. All the rest have disappeared, done in by fires or earthquakes.

Laughter and numbers are like the pyramids in their expected longevity. And that, as I have already said, is a good thing, because they lie at the core, respectively, of humor and mathematics, and these make life bearable for the nobler spirits among us. Bertrand Russell recounts in his autobiography that as an unhappy adolescent he frequently contemplated suicide. But he did not go through with it, he tells us, "because I wished to know more of mathematics." Woody Allen's character in the film *Hannah and Her Sisters* is similarly given to suicidal thoughts, but he is pulled back from the brink when he goes to a revival cinema and sees the Marx Brothers, in *Duck Soup*, playing on the helmets of the soldiers of Freedonia like a xylophone. If our descendants in the Year Million are to find existence worth the bother, they had better have laughter and mathematics.

But what will their mathematics look like? And what will make them laugh?

The first question might seem the easier to answer. Mathematics, after all, is supposed to be the most universal part of human civilization. All terrestrial cultures count, so all terrestrial cultures have number. If there is intelligent life elsewhere in the cosmos, we would expect the same. The one earmark of civilization that is likely to be recognized across the universe is number. In Carl Sagan's science-fiction novel

Contact, aliens in the vicinity of the star Vega beam a series of prime numbers toward earth. The book's heroine (played by Jodie Foster in the movie version of *Contact*) works for SETI (Search for Extraterrestrial Intelligence). She realizes, with a frisson, that the prime-number pulses her radio telescope is picking up must have been generated by some form of intelligent life.

But what if aliens beamed jokes at us instead of numbers? We probably wouldn't be able to distinguish the jokes from the background noise. Indeed, we can barely distinguish the jokes in Shakespeare's plays from the background noise. (Seriously, have you ever laughed during a Shakespeare play?) Just as nothing is more timeless than number, the core of mathematics, nothing is more parochial and ephemeral than humor, the core of laughter. So, at least, we imagine. We are quite confident that a civilization a million years more advanced than our own would find our concept of number intelligible, and we theirs. But their jokes would have us scratching our heads in puzzlement, and vice versa. As for all the bits of culture in between—ranging from (at the more parochial end) literature and art to (at the most universal end) philosophy and physics—well, who knows?

That's how we see matters at the moment. In the Year Million, though, I think the perspective will be precisely the reverse. Humor will be esteemed as the most universal aspect of culture. And number will have lost its transcendental reputation and be looked upon as a local artifact, like a computer operating system or an accounting scheme. If I am right, then SETI scientists should be listening not for prime numbers or the digits of π but for something quite different.

Return to numbers for a moment. In 1907, when he was in his thirties, Bertrand Russell penned a gushing tribute to the glories of mathematics. "Rightly viewed," Russell wrote, mathematics "possesses not only truth, but supreme beauty—a beauty cold and austere, like that of sculpture, without appeal to any part of our weaker nature, without the gorgeous trappings of painting or music, yet sublimely pure, and capable of a stern perfection such as only the greatest art can show." These lines, which play up the transcendent image of mathematics, are often quoted in mathematical popularizations. What one seldom encounters in such books, however, is the rather different view that Russell expressed in his late eighties, when he dismissed his (relatively) youthful rhapsodizing as "largely nonsense." Mathematics, the aged Russell wrote, "has ceased to seem to me non-human in its subject-matter. I have come to believe, though very reluctantly, that it consists of tautologies. I fear that, to a mind of sufficient intellectual power, the whole of mathematics would appear trivial, as trivial as the statement that a four-footed animal is an animal." So, in the course of his life, Russell underwent an evolution in his thinking about mathematics. I think our civilization will have undergone a similar evolution by the Year Million. (Yes, Virginia, phylogeny sometimes recapitulates ontogeny.) Our descendants will view mathematics merely as an elaborate network of tautologies, of strictly local import, one that has proved convenient to us as a bookkeeping scheme for coping with the world.

If mathematics is essentially trivial, its triviality ought to be most apparent at the elementary level, before the smoke and mirrors of higher theory have had a chance to do their work. So let's look at this level. The most fundamental objects in mathematics, everyone would agree, are the counting numbers: 1, 2, 3, and so on. Among such numbers, the prime numbers—those that the aliens were beaming in *Contact*—are supposed to be special. A prime is a number that cannot be split up into smaller factors. (Another way of putting this is to say that a prime is a number that can be evenly divided only by itself and one.) The first few primes are 2, 3, 5, 7, 11, 13, 17, 19, 23, 29, 31, 37 . . . The primes are the atoms of arithmetic: all the rest of the numbers, called "composite" numbers, can be built up by multiplying primes together in various combinations. Thus the number 666 can be obtained by multiplying $2 \times 3 \times 3 \times 37$. One can prove, with just a little trouble, that every composite number can be put together in one and only one way as a product of prime numbers. This is often called the fundamental theorem of arithmetic.

So far so good: everything looks tautological enough. Then let's move to the next obvious question: How many prime numbers are there? This question was posed by Euclid in the third century B.C.E., and the answer is contained in proposition 20 of his *Elements*: there are infinitely many primes. Euclid's proof of this proposition is perhaps the first truly elegant bit of reasoning in the history of mathematics. It can fit into a single sentence: If there were only finitely many primes, then, by multiplying them all together and adding 1, you would get a new number that could not be divided by any prime at all, which is impossible. (This new number would leave a remainder of 1 if it was divided by any of the numbers on the supposedly finite list of primes; so it would have to be either a prime number itself or divisible by some prime that was not on the original list. In either case, the original finite list of primes must be incomplete. So no finite list can encompass all the primes. Therefore there must be infinitely many of them.)

Once we know that the primes go on forever, the next question that naturally arises is this: How are these atoms of arithmetic scattered among the rest of the numbers? Is there a pattern? Primes turn up rather frequently among the smaller numbers, but they get scarcer as you move through the number sequence. Of the first ten numbers, four are prime (2, 3, 5, and 7). Of the first one hundred numbers, twenty-five are prime. Jumping on a bit, of the numbers between 9,999,900 and 10,000,000, nine are prime; among the next hundred numbers, from 10,000,000 to 10,000,100, only two are prime (10,000,019 and 10,000,079). It is possible to find stretches of numbers as long as you please that are completely prime-free. But there are also very large primes that clump together, like 1,000,000,009,649 and 1,000,000,009,651. (Primes differing only by two in this way are called twin primes; it is an open question whether there are infinitely many of them.) Prime numbers seem to crop up almost at random, sprouting like weeds among the rest of the numbers. "There is no apparent reason why one number is a prime and another not," declared the mathematician Don Zagier in his inaugural lecture at Bonn University in 1975. "To the contrary, upon looking at these numbers one has the feeling of being in the presence of one of the inexplicable secrets of creation."

Despite their simple definition, the primes appear to have a complex and timeless reality all their own, one quite independent of our minds. They are transcendently mysterious in a way that Russell's proposition "a four-footed animal is an animal" is not. But are they completely lawless? That would be very surprising, given their role as the building blocks of arithmetic. And in fact they do obey a law. But to find this law, strangely, one must ascend many stories in the edifice of mathematics: from the humble counting numbers through the integers, the fractions, and the real numbers, all the way to complex numbers with "imaginary" parts. (Historically, that ascent took longer than two millennia.) And then, at that lofty level, one runs into a conundrum known as the Riemann zeta hypothesis.

■

By the near-unanimous judgment of mathematicians, the Riemann zeta hypothesis is the greatest unsolved problem in all of mathematics. It may be the most difficult problem ever conceived by the human mind. The Riemann in question is Bernhard Riemann, a nineteenth-century German mathematician. "Zeta" refers to the zeta function, a creature of higher mathematics that, as Riemann was the first to realize, holds the secret of the primes. In 1859, in a brief but exceedingly profound paper, Riemann put forward a hypothesis about the zeta function. If his hypothesis is true, then there is a hidden harmony to the primes, one that is rather beautiful. If it is false, then the music of the primes could turn out to be somewhat ugly, like that produced by an orchestra out of balance.

Which will it be? For the last century and a half, mathematicians have been striving in vain to resolve the Riemann zeta hypothesis. In a celebrated speech delivered in 1900 before an international conference in Paris, David Hilbert included it on his list of the twenty-three most important problems in mathematics. (He later declared it to be the most important "not only in mathematics, but absolutely the most important.") The Riemann hypothesis was the only problem on Hilbert's list to survive the century undispatched. In 2000, on the one hundredth anniversary of Hilbert's speech, a small group of the world's leading mathematicians held a press conference at the Collège de France to announce a fresh set of seven "Millennium Prize Problems," the solution of any of which would be rewarded with a prize of one million dollars. (The prize money is courtesy of the Clay Mathematics Institute, founded by the Boston investor Landon T. Clay.) To no one's surprise, the Riemann hypothesis made this list too.

The Riemann zeta hypothesis is more than just the key to understanding the primes. So central is it to mathematical progress that its truth has simply been assumed—perhaps rashly—in the provisional proofs of thousands of theorems (which are said to be "conditioned" on the hypothesis). If it turns out to be false, the part of higher mathematics that is built upon it will collapse. (Fermat's last theorem, which was proved in 1995, played no such structural role in mathematics and hence was far less important.)

The zeta function, fittingly, has its origins in music. If you pluck a violin string, it vibrates to create not only the note to which it is tuned but also all possible overtones. Mathematically, this combination of sounds corresponds to the infinite sum $1 + \frac{1}{2} + \frac{1}{3} + \frac{1}{4} + \ldots$, which is known as the harmonic series. If you take every term in this series and raise it to the variable power s, you get the zeta function:

$$\zeta(s) = 1 + \left(\frac{1}{2}\right)^s + \left(\frac{1}{3}\right)^s + \left(\frac{1}{4}\right)^s + \ldots$$

This function was introduced around 1740 by Leonhard Euler, who proceeded to make a striking discovery. He found that the zeta function, an infinite *sum* running though *all* the numbers, could be rewritten as an infinite *product* running through *just the primes* (which appear as reciprocals):

$$\zeta(s) = \frac{1}{1 - \left(\frac{1}{2}\right)^s} \times \frac{1}{1 - \left(\frac{1}{3}\right)^s} \times \frac{1}{1 - \left(\frac{1}{5}\right)^s} \times \frac{1}{1 - \left(\frac{1}{7}\right)^s} \times \frac{1}{1 - \left(\frac{1}{11}\right)^s} \times \ldots$$

Although Euler was the greatest mathematician of his time, he did not fully grasp the potential of the infinite-product formula he had discovered. "Mathematicians have tried in vain to this day to discover some order in the sequence of prime numbers," Euler wrote, "and we have reason to believe that it is a mystery into which the human mind will never penetrate."

Half a century later, Carl Friedrich Gauss made the first real breakthrough since Euclid in understanding the prime numbers. As a boy, Gauss enjoyed tallying up how many primes there were in each block of a thousand numbers. Such computations were a good way to beguile "an idle quarter of an hour," he wrote to a friend, "but at last I gave it up without quite getting through a million." In 1792, when he was fifteen years old, Gauss noticed something interesting. Although the primes cropped up seemingly at random, there did appear to be some regularity to their overall flow. A good estimate for how many primes there are up to a given number could be obtained by dividing that number by its natural logarithm. Suppose, for example, you want to know how many primes there are up to a million. Take out your pocket calculator, punch in 1,000,000, and divide it by the ln(1,000,000). Out pops 72,382. The actual number of primes up to a million is 78,498, so the estimate is off by about 8 percent. But the percentage error heads toward zero as the numbers get bigger.

What Gauss had discovered was "the coin that Nature had tossed to choose the primes" (in the words of the British mathematician Marcus du Sautoy). It was uncanny that this coin should be weighted by the natural logarithm, which arose in the continuous world of the calculus and would seem wholly unrelated to the chunky world of the counting numbers. (The logarithmic function is defined by the area under a certain curve.) Gauss could not prove that the natural log function would continue to describe the waning of the primes through the entire infinity of numbers; he was just making an empirical guess. Nor could he explain its inexactness: why it failed to say *precisely* where the next prime would turn up.

It was Riemann who saw through the illusion of randomness. In 1859, in a paper less than ten pages long, he made a series of moves that cracked the mystery of the primes. He began with the zeta function. Euler had seen this function as ranging only over "real" values. (The real numbers, which correspond to the points on a line, comprise the whole numbers, both positive and negative; rational numbers, which can be represented by fractions; and irrational numbers, like π or e, which can be represented by non-repeating decimals.) But Riemann ventured beyond Euler, enlarging the zeta function to take in the *complex* numbers.

Complex numbers have two distinct parts, one "real" and one "imaginary." (The "imaginary" part involves $\sqrt{-1}$. One typical complex number is $2 + 3\sqrt{-1}$, with 2 being the real part and $3\sqrt{-1}$ the imaginary part.) Since complex numbers have two parts, they can be thought of as two-dimensional; instead of forming a line (like the real numbers), they form a plane. Riemann decided to extend the zeta function over this complex plane. At every point in the complex plane, he showed, the zeta function determines an altitude. It thus gives rise to a vast abstract landscape consisting of mountains, hills, and valleys that stretch forever in every direction—the zeta landscape. The most interesting points in the zeta landscape, he found, are the ones with zero altitude—that is, the sea-level points. These points are called the zeros of the zeta function, because they correspond to those complex numbers that, if plugged into the zeta function, yield the output zero. Using these complex "zeros" of the zeta function—of which there are infinitely many in the zeta landscape—Riemann was able to do a marvelous thing: he produced, for the first time ever, a formula that described *exactly* how the infinity of primes arranged themselves in the number sequence.

This discovery opened up a metaphorical dialogue between mathematics and music. Before Riemann, only random noise could be heard in the primes. Now there was a new way to listen to their music. Each zero of the zeta function, when fed into Riemann's prime formula, produces a wave resembling a pure musical tone. When these pure tones are all combined, they generate the harmonic structure of the prime numbers. Riemann found that the location of a given zero point in the zeta landscape determines the pitch and volume of its corresponding musical note. The farther north the zero is, the higher the pitch. And—more important—the farther east it is, the greater the loudness. Only if all the zero points lie in a fairly narrow longitudinal strip of the zeta landscape will the orchestra of the primes be in balance, with no instrument drowning out the others. But Riemann went further. After navigating just a tiny part of the infinite zeta landscape, he boldly asserted that all its zeros were precisely arrayed along a "critical line" running from south to north. And this is the claim that subsequently became known as the Riemann zeta hypothesis.

"If Riemann's Hypothesis is true," du Sautoy has written, "it will explain why we see no strong patterns in the primes. A pattern would correspond to one instrument playing louder than the others. It is as if each instrument plays its own pattern, but by combining together so perfectly, the patterns cancel themselves out, leaving just the formless ebb and flow of the primes." There is something magical in the way the infinity of zero points in the zeta landscape collectively control how the infinity of primes occur among the counting numbers: the more regimented the zeros are on one side of the looking glass, the more random the primes appear on the other.

But are the zeros as perfectly regimented as Riemann believed? If the Riemann hypothesis is false, a single zero off the critical line would suffice to refute it. Calculating where these zeros lie is no trivial matter. Riemann's own navigation of the zeta landscape revealed that the first few sea-level points lined up the way he expected they would. In the early twentieth century, hundreds more zeros were computed by hand. Since then, computers have located billions of zeros, and every one of them lies precisely on the critical line. One might think that the failure thus far to find a counterexample to the Riemann hypothesis increases the odds that it is true. This is a matter of some controversy. There are, after all, infinitely many zero points of the zeta function, and it may be that they only reveal their true colors in unimaginably distant reaches of the zeta landscape—reaches whose exploration may lie well beyond the Year Million. Those who blithely assume the truth of the Riemann conjecture should keep in mind an interesting pattern in the history of mathematics: whereas long-standing conjectures in algebra (like Fermat's theorem) typically turn out to be true, long-standing conjectures in analysis (like the Riemann conjecture) often turn out to be false.

The majority of mathematicians today who cleave to the Riemann hypothesis do so primarily on aesthetic grounds: it is simpler and more beautiful than its negation, and it leads to the most "natural" distribution of primes. "If there *are* lots of zeros off the line—and there might be—the whole picture is just horrible, horrible, very ugly," the mathematician Steve Gonek has said. The hypothesis is unlikely to have any practical consequences, but that is of little import to the mathematicians who pursue it. "I have never done anything 'useful,'" G. H. Hardy bragged in his famous book *A Mathematician's Apology*. "No discovery of mine has made, or is likely to make, directly or indirectly, for good or ill, the least difference to the amenity of the world." Mathematicians like Hardy acknowledge two motives: one is the sheer pleasure of doing mathematics. The other is the sense that they are like astronomers peering out at a Platonic cosmos of numbers, a cosmos that transcends human culture and that of any other civilizations there might be, either now or in the future. Hardy adds that "317 is a prime, not because we think so, or because our minds are shaped in one way rather than another, but *because it is so*, because mathematical reality is built that way." Alain Connes, a French mathematician who is widely deemed a leading candidate to prove the Riemann hypothesis, is also unabashed in his Platonism. "For me," Connes has said, "the sequence of prime numbers . . . has a reality that is far more permanent than the physical reality surrounding us."

But will this still seem true in the Year Million? I think that prime numbers will lose their transcendental reputation when we come to understand them more completely. Then we will see that like the rest of mathematics (or like religion, for that matter) they are man-made, a terrestrial artifact. And when can we expect the great devalorization? Paul Erdős, the most prolific (and peripatetic) of modern mathematicians, is reputed to have declared, "It will be another million years, at least, before we understand the primes." The Copernican principle yields a rather different estimate. The Riemann zeta conjecture has been open since it was first posed by Riemann himself around 160 years ago. That means we can be 95 percent certain that it will survive as an open problem for another four years or so (1/39 × 160) but that it will be dispatched within the next six millennia (39 × 160)—well short of the Year Million. If and when it is dispatched, the prime numbers will finally be stripped of their cosmic otherness.

The prime numbers define the zeta function; the zeta function determines the zero points; and the zero points collectively harbor the secrets of the primes. Resolving the Riemann zeta hypothesis will close this tight little circle, rendering the "mystery" of the primes as tautological as the statement "a four-footed animal is an animal." My prediction, then, is that long before the Year Million, mathematicians will have awakened from their collective Platonist dream. No one will give a thought to beaming prime numbers throughout the cosmos. Our descendants will dismiss them just as the hero of Bertrand Russell's story "The Mathematician's Nightmare" did—by saying, "Avaunt! You are only Symbolic Conveniences!"

■

And how about laughter? As I observed earlier, nothing is thought to be more local, parochial, and ephemeral than the sort of "humor" that evokes a chuckle. Or more lowly. For much of human history, the comical has been a mix of lewdness, aggression, and mockery. As for the peculiar panting and chest-heaving behavior to which it gives rise, that was viewed as a "luxury reflex" serving no obvious adaptive purpose.

In recent years, though, practitioners of the fanciful art of evolutionary psychology have been more inventive in coming up with Darwinian rationales for laughter. One of the most plausible is from the neuroscientist V. S. Ramachandran. In his 1998 book, *Phantoms in the Brain* (written with Sandra Blakeslee), Ramachandran advanced what might be called the "false alarm" theory of laughter. A seemingly threatening situation presents itself; you go into the fight-or-flight mode; the threat proves spurious; you alert your (genetically close-knit) social group to the absence of actual danger by emitting a stereotyped vocalization—one that is amplified as it passes, contagiously, from member to member.

Once this mechanism was put in place by evolution, the theory goes, it could be hijacked for other purposes—expressing hostility toward (and superiority vis-à-vis) other social groups, for instance, or ventilating forbidden sexual impulses within one's own group. But at the core of the original "false alarm" mechanism of laughter is *incongruity*: the incongruity of a grave threat revealing itself to be trivial; of a minatory "something" evaporating into a harmless "nothing." And in the evolution of humor over the millennia, the perception of incongruity has played a more and more dominant role. Laughter, at its highest, is now regarded as the expression of an intellectual emotion. Indeed, the upper bound of the evolution of jocularity may be the Jewish joke, where a Talmudic playfulness toward language and logic reigns. (Think of your favorite Groucho Marx or Woody Allen line.) On this intellectualist view, the greatest stimulus to laughter is pure, abstract incongruity. Every good joke, as Schopenhauer held, is a disrupted syllogism. (For example, "The important thing is sincerity. If you can fake that, you've got it made.") And incongruity is the opposite of boring old tautology. And just as universal.

That is why I think humor and mathematics will have changed places by the Year Million. But what will jokes look like in that distant epoch? The higher laughter is called forth when we see an incongruity resolved in some clever way, resulting in an emotional shudder of pleased recognition. We imagined we were apprehending something odd and mysterious, but suddenly we find ourselves holding nothing at all. The Riemann zeta hypothesis, when it is finally dispatched in the aeons to come, will provide just such a resolution. Amid peals of laughter, the Platonic otherness of the primes will dissolve into trivial tautology. It is sobering to think that what is today regarded as the hardest problem ever conceived by the human mind may well be, in the Year Million, a somewhat broad joke, fit for schoolchildren.

5

Sir Francis Galton, the Father of Statistics . . . and Eugenics

In the 1880s, residents of cities across Britain might have noticed an aged, bald, bewhiskered gentleman sedulously eyeing each and every girl he passed on the street while manipulating something in his pocket. What they were seeing was not lechery in action but science. Concealed in the man's pocket was a device he called a "pricker," which consisted of a needle mounted on a thimble and a cross-shaped piece of paper. By pricking holes in different parts of the paper, he could surreptitiously record his rating of a female passerby's looks, on a scale ranging from attractive to repellent. After many months of wielding his pricker and tallying results, he drew a "beauty map" of the British Isles. London proved the epicenter of beauty, Aberdeen of its opposite.

Such research was entirely congenial to Francis Galton, a man who took as his motto, "Whatever you can, count." Galton was one of the great Victorian innovators—not quite so great, perhaps, as his cousin Charles Darwin, but certainly more versatile. He explored unknown regions of Africa. He pioneered the fields of weather forecasting and fingerprinting. He discovered statistical ideas that revolutionized the methodology of science. Personally, Galton was an attractive and clubbable fellow, if a bit of a snob. Yet today he is most remembered for an achievement that puts him in a decidedly sinister light: he was the father of eugenics, the science, or pseudoscience, of "improving" the human race by selective breeding.

Eugenics, as would appear obvious from its subsequent career, is an evil thing to have fathered. Enthusiasm for Galton's ideas led to the forcible sterilization of hundreds of thousands of people in the United States and Europe deemed genetically unfit and contributed to Nazi racial policies that culminated in the Holocaust. Today, most of us can see that the notion of uplifting humanity by getting "desirables" to breed more and "undesirables" to breed less was misbegotten from the start, both on scientific and on ethical grounds. We look back at its progenitor with a slight shudder of revulsion and chalk it up to moral progress that we have put the Galtonian gospel behind us. But perhaps we should not be so smug. In the new age of genetic engineering, it is becoming apparent that the eugenic temptation has not gone away at all; it is simply taking a new form, one that may prove harder to resist. If Galton, as our image of him would have it, was a talented and fundamentally decent man seduced by an evil concept, it might be of more than historical interest to reconsider how he went wrong.

Francis Galton "came into the world in the wake of Wellington and Waterloo, and went out at the dawn of motor cars and aeroplanes," in the words of his biographer Martin Brookes. Born in 1822 into a wealthy and distinguished Quaker family—his maternal grandfather was Erasmus Darwin, a revered physician and botanist who wrote poetry about the sex lives of plants—Galton enjoyed a pampered upbringing. As a child, he reveled in his own precocity: "I am four years old and can read any English book. I can say all the Latin substantives and adjectives and active verbs besides 52 lines of Latin poetry. I can cast up any sum in addition and multiply by 2, 3, 4, 5, 6, 7, 8, 10. I can also say the pence table. I read French a little and I know the Clock." When Galton was sixteen, his father decided that he should follow a medical career, like his distinguished grandfather. He was sent to train in a hospital, but the screams of the patients on the operating table in those pre-anesthesia days proved off-putting. Seeking guidance from his cousin Charles Darwin, who was just back from his trip around the world on the HMS *Beagle*, Galton was advised to "read Mathematics like a house on fire." So he enrolled at Cambridge, where, despite his invention of a "gumption-reviver machine" to drip water on the flagging scholar's head, he promptly suffered a breakdown from overwork.

This pattern of frantic intellectual activity followed by nervous collapse was to continue throughout Galton's life. His need to earn a living, though, ended with the death of his father when Galton was twenty-two. Now in possession of a handsome inheritance and liberated from the burden of paternal expectations, he took up a life of sporting hedonism. In 1845, he went on a hippo-shooting expedition down the Nile (he proved a woeful shot), then trekked by camel across the Nubian Desert. Continuing his tour into the Near East, he taught himself Arabic and apparently caught a venereal disease from a prostitute—which may account for a noticeable cooling in the young man's ardor for women.

The world at that time still contained vast uncharted areas, and exploring them seemed an apt vocation to this rich Victorian bachelor. In 1850, Galton sailed to southern Africa and organized an expedition into parts of the interior never before seen by a white man. Before setting out, he purchased a theatrical crown in Drury Lane that he planned to place "on the head of the greatest or most distant potentate I should meet with." Improvising survival skills as he pursued his thousand-mile journey through the bush, Galton contended with searing heat, scarce water, tribal warfare, marauding lions that consumed his mules and horses, shattered axles, dodgy guides, and native helpers whose conflicting dietary superstitions made it impossible to settle on a mutually agreeable meal from the caravan's mobile larder of sheep and oxen. Methodical in his observations, he became adept in the use of the sextant, at one point employing this navigational device to measure from afar the curves of an especially buxom native woman—"a Venus among Hottentots."

The climax of the journey was his encounter with King Nangoro, a tribal ruler locally reputed to be the fattest man in the world. Nangoro was fascinated by the Englishman's white skin and straight hair and moderately pleased when the tacky stage crown was placed on his head. But Galton committed an irreparable gaffe. When the king dispatched his niece, smeared in butter and red ocher, to his guest's tent to serve as a wife for the night, Galton, wearing his one clean suit of white linen, found the naked princess "as capable of leaving a mark on anything she touched as a well-inked printer's roll . . . so I had her ejected with scant ceremony."

Galton's feats made him famous: on his return to England, the thirty-year-old explorer was celebrated in the newspapers and awarded a gold medal by the Royal Geographical Society. After writing a bestselling book on how to survive in the African bush, he decided that he had had enough of the adventurer's life. Still a pleasant-looking man despite his increasing baldness—for which he attempted to compensate by growing a pair of enormous sideburns—he married a rather plain woman from an intellectually distinguished family, with whom he never succeeded in having children (possibly because venereal disease had left him sterile). Purchasing a mansion in South Kensington conveniently near the many clubs and societies to which he now belonged, he settled down to a life of scientific dilettantism. His true métier, he had always felt, was measurement. In pursuit of it, he conducted elaborate experiments in the science of tea making, deriving equations for brewing the perfect cup. He set about calculating the total volume of gold in the world, concluding, to his astonishment, that it was considerably less than the volume of his dining room.

Eventually, his interest hit on something that was actually important: the weather. Meteorology could barely be called a science in those days; the forecasting efforts of the British government's first chief weatherman met with such public ridicule that he ended up slitting his throat. Taking the initiative, Galton solicited reports of conditions all over Europe and then created the prototype of the modern weather map. He also discovered an important new weather pattern he called the "anti-cyclone"—better known today as the high-pressure system.

Galton might have puttered along for the rest of his life as a minor gentleman-scientist had it not been for a dramatic event: the publication of Darwin's *Origin of Species* in 1859. Reading his cousin's book filled Galton with a sense of clarity and purpose. One thing in it struck him with special force: to illustrate how natural selection shaped species, Darwin cited the breeding of domesticated plants and animals by farmers to produce better strains. Perhaps, Galton dreamed, human evolution could be deliberately guided in the same way. "If a twentieth part of the cost and pains were spent in measures for the improvement of the human race that is spent on the improvements of the breed of horses and cattle, what a galaxy of genius might we not create!" he wrote in an 1864 magazine article, his opening eugenic salvo. (It was two decades later that he coined the actual word "eugenics," from the Greek for "wellborn.")

Leaping beyond Darwin, who had thought mainly about the evolution of physical features, like wings and eyes, Galton applied the same hereditary logic to mental traits, like talent and virtue. That put him in opposition to the philosophical orthodoxy represented by John Locke, David Hume, and John Stuart Mill, who had maintained that the mind was a blank slate, to be written upon by experience. "I have no patience," Galton wrote, "with the hypothesis occasionally expressed, and most often implied, especially in tales written to teach children to be good, that babies are born pretty much alike and that the sole agencies in creating differences between boy and boy, and man and man, are steady application and moral effort."

It was Galton who coined the phrase "nature versus nurture," which still reverberates in debates today. (It was probably suggested by Shakespeare's *Tempest*, in which Prospero laments that his adopted son, Caliban, is "A devil, a born devil, on whose nature / Nurture can never stick.") What made him so sure that nature dominated nurture in determining a person's talent and temperament? The idea first arose in his mind at Cambridge, where he noticed that the top students had relatives who had also excelled there; surely, he reasoned, such runs of family success were not a matter of mere chance. His hunch was strengthened during his travels, which gave him a vivid sense of what he called "the mental peculiarities of different races."

Galton made an honest effort to justify his belief in nature over nurture with hard evidence. In his 1869 book, *Hereditary Genius*, he assembled long lists of "eminent" men—judges, poets, scientists, even oarsmen and wrestlers—to show that excellence ran in families. To counter the objection that social advantages, rather than biology, might be behind this, he used the adopted sons of popes as a kind of control group. His case that mental ability was largely hereditary elicited skeptical reviews, but it impressed Darwin. "You have made a convert of an opponent in one sense," he wrote to Galton, "for I have always maintained that, excepting fools, men did not differ much in intellect, only in zeal and hard work." But Galton's work had hardly begun. If his eugenic utopia was to be a practical possibility, he needed to know more about how heredity worked; in the absence of such knowledge, even the strictest oversight of marriage and reproduction might fail to bring about the hoped-for betterment of humanity. Galton's belief in eugenics thus led him to try to discover the law of inheritance. And that, in turn, led him to statistics.

Statistics at that time was a dreary thing, a welter of population numbers, trade figures, and the like. It was devoid of mathematical interest, save for a single concept: the bell curve. Oddly enough, the bell curve—also known as the normal or Gaussian distribution (after Carl Friedrich Gauss, one of its multiple discoverers)—first arose in astronomy. In the eighteenth century, astronomers noticed that the errors in their measurements of the positions of planets tended to fall into a distinctive pattern. The measurements clustered symmetrically around the true value, most of them quite close to it, a few farther away on either side. When graphed, the distribution of errors had the shape of a bell. In the early nineteenth century, a Belgian astronomer named Adolphe Quetelet observed that the same "law of error" that arose in astronomy also applied to many human phenomena. Gathering information on the chest sizes of five thousand Scottish soldiers, for example, Quetelet found that data traced out a bell curve centered on the average chest size, about forty inches.

Why is the bell curve so ubiquitous? Mathematics yields the answer. It is guaranteed to arise whenever some variable (like human height) is determined by lots of little causes (genes, diet, health, and so on) operating more or less independently. For Quetelet, the bell curve represented accidental deviations from a sort of Platonic ideal he called *l'homme moyen*—the average man. When Galton stumbled upon Quetelet's work, however, he exultantly saw the bell curve in a new light: what it described were not accidents to be overlooked but differences that reveal the variability on which evolution depends. His quest to find the laws that governed how these differences were transmitted from one generation to the next led to what has been justly deemed Galton's greatest gifts to science: regression and correlation.

Although Galton's chief interest was the inheritance of mental abilities, such as intelligence, he knew they would be hard to measure. So he focused on physical traits, such as height. The only rule of heredity known at the time was the vague "Like begets like." Taller parents tend to have taller children; shorter parents tend to have shorter children. But individual cases were unpredictable. Hoping to find some larger pattern, Galton set up an "anthropometric laboratory" in London in 1884. Drawn by his fame, thousands of people streamed in and obligingly submitted to measurement of their height, weight, reaction time, pulling strength, color perception, and so on. Among the visitors was William Gladstone, the prime minister at the time. "Mr. Gladstone was amusingly insistent about the size of his head . . . but after all it was not so very large in circumference," noted Galton, who took pride in his own massive bald dome.

After obtaining height data from 205 pairs of parents and 928 of their adult children, Galton plotted the points on a graph, with the parents' heights represented on one axis and the children's on the other. He then penciled a straight line through the cloud of points to capture the trend it represented. The slope of this line turned out to be two-thirds. What this meant, Galton realized, was that exceptionally tall (or short) parents had children who, on average, were only two-thirds as exceptional as they were. In other words, when it came to height, children tended to be more mediocre than their parents. The same, he had noticed years earlier, seemed to be true in the case of "eminence": the children of J. S. Bach, for example, might have been more musically distinguished than average, but they were less distinguished than their father. Galton called this phenomenon "regression toward mediocrity." Regression analysis furnished a way of predicting one thing (like a child's height) from another (his parents') when the two things were fuzzily related. Galton went on to develop a measure of the *strength* of such fuzzy relationships, one that could be applied even when the things related were different in kind—like rainfall and crop yield, or cigarette consumption and lung cancer, or class size and academic achievement. He called this more general technique "correlation."

The result was a major conceptual breakthrough. Until then, science had pretty much been limited to deterministic laws of cause and effect. Such laws are not easy to find in the biological world, where multiple causes often blend together in a messy way. Thanks to Galton, statistical laws began to gain respectability in the scientific world. His discovery of regression toward mediocrity—or regression to the mean, as it is now called—has resonated even more widely. In his 1996 book, *Against the Gods: The Remarkable Story of Risk*, Peter L. Bernstein wrote, "Regression to the mean motivates almost every variety of risk-taking and forecasting. It is at the root of homilies like 'What goes up must come down,' 'Pride goeth before a fall,' and 'From shirtsleeves to shirtsleeves in three generations.'"

Yet, as straightforward as it seems, the idea of regression has been a snare for the sophisticated and the simple alike. The most common misconception is that it implies convergence over time. If very tall parents tend to have somewhat shorter children, and very short parents tend to have somewhat taller children, doesn't that mean that eventually everyone should be the same height? No, because regression works backward in time as well as forward: very tall children tend to have somewhat shorter parents, and very short children tend to have somewhat taller parents.

The key to understanding this seeming paradox is to realize that regression to the mean arises when enduring factors, which might be called skill, mix causally with transient factors, which might be called luck. Consider the case of sports, where regression to the mean is often mistaken for choking or slumping. Major League Baseball players who managed to bat over .300 last season did so through a combination of skill and luck. Some of them are truly great players who had a so-so year, but the great majority are merely good players who had a lucky year. There is no reason why the latter group should be equally lucky next year; that is why around 80 percent of them will see their batting averages decline.

To mistake regression for a real force that causes talent or quality to dissipate over time, as so many have, is to commit what has been called Galton's fallacy. In 1933, a Northwestern University professor named Horace Secrist produced a book-length example of the fallacy in *The Triumph of Mediocrity in Business*, in which he argued that since highly profitable firms tend to become less profitable and highly unprofitable ones tend to become less unprofitable, all firms will soon be mediocre. A few decades ago, the Israeli Air Force came to the conclusion that blame must be more effective than praise in motivating pilots, because poorly performing pilots who were blamed made better landings subsequently, whereas high performers who were praised made worse ones. (It is a sobering thought that we might generally tend to overrate censure and underrate praise because of the regression fallacy.) In 1990, an editorialist for *The New York Times* erroneously argued that the regression effect alone would ensure that racial differences in IQ would disappear over time.

Could Galton himself have committed Galton's fallacy? Martin Brookes, in his 2004 biography of Galton, *Extreme Measures*, insists that he did. "Galton completely misread his results on regression," Brookes writes. "Human heights do not have a tendency to become more average with each generation. But regression, Galton believed, proved that they did." Even worse, Brookes claims, Galton's muddleheadedness about regression led him to reject the Darwinian view of evolution and to adopt a more extreme and unsavory version of eugenics. Suppose regression really did act as a sort of gravity, always pulling individuals back toward the population average. Then it would seem to follow that evolution could not take place through a gradual series of small changes, as Darwin envisaged. It would require large, discontinuous changes that are somehow immune to regression to the mean. Such leaps, Galton thought, would result in the appearance of strikingly novel organisms, or "sports of nature," that would shift the entire bell curve of ability. And if eugenics was to have any chance of success, it would have to work the same way as evolution. In other words, these sports of nature would have to be enlisted as studs to create a new breed. Only then could regression be overcome and progress made.

But Galton was not quite so confused as his biographer Brookes makes him out to be. It took Galton nearly two decades to work out the subtleties of regression, an achievement that, according to Stephen M. Stigler, a historian at the University of Chicago, "should rank with the greatest individual events in the history of science—at a level with William Harvey's discovery of the circulation of the blood and with Isaac Newton's of the separation of light." By 1889, when he published his most influential book, *Natural Inheritance*, his grasp of it was nearly complete. He knew that regression had nothing special to do with life or heredity. He knew that it was independent of the passage of time. (Regression to the mean even held between brothers, he observed; exceptionally tall men tend to have brothers who are somewhat less tall.) In fact, as Galton was able to show by a neat geometric argument, regression is a matter of pure mathematics, not an empirical force. Lest there be any doubt, he disguised the case of hereditary height as a problem in mechanics and sent it to a trained mathematician at Cambridge, who, to Galton's delight, confirmed his finding. "I never felt such a glow of loyalty and respect towards the sovereignty and magnificent sway of mathematical analysis as when his answer reached me," Galton wrote.

Even as he laid the foundations for the statistical study of human heredity, Galton continued to pursue many other intellectual interests, some important, some merely eccentric. He invented a pair of submarine spectacles that permitted him to read while submerged in his bath and a bicycle speedometer that (as Brookes describes it) consisted of "nothing more than an egg-timer, which the cyclist was supposed to hold while counting the revolution of the pedals." His scientific papers, numbering more than three hundred, included "Arithmetic by Smell" and "Strawberry Cure for Gout." He stirred up controversy by using statistics to investigate the efficacy of prayer. (Petitions to God, he concluded, were powerless to protect people from sickness.) He tried, apparently with some success, to induce in himself a state of temporary insanity by walking through Piccadilly imagining that every person and object around him was a spy. Prompted by a near approach of the planet Mars to Earth, he devised a celestial signaling system to permit communication with Martians. More usefully, he put the nascent practice of fingerprinting on a rigorous basis by classifying patterns and proving that no two fingerprints were exactly the same,—a great step forward for Victorian policing. He created the first psychological questionnaire, which he distributed to scientists to learn about their powers of mental imagery. He also invented the technique of word association, which he used to plumb his own unconscious, decades before Freud.

Galton remained restlessly active through the turn of the century. In 1900, eugenics received a big boost in its prestige when Gregor Mendel's work on the heredity of peas came to light. Suddenly hereditary determinism was the scientific fashion. Although Galton was plagued by deafness and asthma (which he tempered by smoking hashish), he gave a major address on eugenics before the Sociological Society in 1904. "What nature does blindly, slowly, and ruthlessly, man may do providently, quickly, and kindly," he declared. An international eugenics movement was springing up, and Galton was hailed as its hero. In 1909, he was honored with a knighthood. Two years later, he died at the age of eighty-eight.

In his long career, Galton did not come close to proving the central axiom of eugenics: that, when it comes to talent and virtue, nature dominates nurture. Yet he never doubted its truth, and many scientists came to share his conviction. Darwin himself, in *The Descent of Man*, wrote, "We now know, through the admirable labours of Mr. Galton, that genius . . . tends to be inherited." Given this axiom, there are two ways of putting eugenics into practice: "positive" eugenics, which means getting superior people to breed more; and "negative" eugenics, which means getting inferior ones to breed less. For the most part, Galton was a positive eugenicist. He stressed the importance of early marriage and high fertility among the genetic elite, fantasizing about lavish state-funded weddings in Westminster Abbey with the queen giving away the bride as an incentive. Always hostile to religion, he railed against the Roman Catholic Church for imposing celibacy on some of its most gifted members over the centuries. (The dysgenic results of priestly celibacy, he thought, could be seen most clearly in the decline of Spain.) He hoped that spreading the eugenic gospel would make the gifted aware of their responsibility to procreate for the good of the human race. But Galton did not believe that eugenics could be entirely an affair of moral suasion. Worried by evidence that the poor in industrial Britain were breeding disproportionately, he urged that charity be redirected away from them and toward the "desirable classes." To prevent "the free propagation of the stock of those who are seriously afflicted by lunacy, feeble-mindedness, habitual criminality, and pauperism," he urged "stern compulsion," which might take the form of marriage restrictions or even sterilization.

Galton's eugenic proposals were benign compared with those of famous contemporaries who rallied to his cause. H. G. Wells, for instance, was an unabashed advocate of negative eugenics, declaring that "it is in the sterilisation of failures, and not in the selection of successes for breeding, that the possibility of an improvement of the human stock lies." George Bernard Shaw championed eugenic sex as an alternative to prescientific procreation through marriage. "What we need," Shaw said, "is freedom for people who have never seen each other before, and never intend to see each other again, to produce children under certain definite public conditions, without loss of honor." Although Galton was a conservative, his creed caught on with progressive figures like Harold Laski, John Maynard Keynes, and Sidney and Beatrice Webb. In the United States, New York disciples founded the Galton Society, which met regularly in the American Museum of Natural History, and popularizers helped the rest of the country become eugenic-minded. "How long are we Americans to be so careful for the pedigree of our pigs and chickens and cattle—and then leave the *ancestry of our children* to chance or to 'blind' sentiment?" asked a placard at an exposition in Philadelphia.

Four years before Galton's death, the Indiana legislature passed the first state sterilization law "to prevent the procreation of confirmed criminals, idiots, imbeciles, and rapists." Soon most of the other states followed its lead. In 1927, Oliver Wendell Holmes Jr., delivering a Supreme Court decision that upheld the legality of Virginia's sterilization law after it had been applied to a young woman whose mother was deemed feebleminded and who had already borne a daughter deemed likewise, declared, "Three generations of imbeciles are enough." All told, there were some sixty thousand court-ordered sterilizations of Americans judged eugenically unfit. Large numbers were also forcibly sterilized in Canada, Sweden, Norway, and Switzerland (although not, notably, in Britain). As the eugenics movement spread throughout the rest of Europe, Latin America, and east to Japan, Galton's program seemed launched "virtually as a planetary revolution," in the words of the historian Daniel J. Kevles.

It was in Germany that eugenics took its most horrific form. Galton's creed had aimed at the uplift of humanity as a whole; although he shared the racial prejudices that were common in the Victorian era, the concept of race did not play much of a role in his theorizing. German eugenics, by contrast, quickly morphed into *Rassenhygiene*—"race hygiene." The Aryan race was in a struggle for domination with inferior races, German race hygienists believed, and its genetic material must not be allowed to deteriorate through the unfettered reproduction of the unfit. Under Hitler, nearly 400,000 people with putatively hereditary conditions like feeblemindedness, alcoholism, and schizophrenia were forcibly sterilized. In time, many were simply murdered. The Nazis also took "positive" eugenic measures; the *Lebensborn* (well of life) program coupled unmarried women who were deemed highly fit with members of the SS in hopes of producing Aryan offspring of the highest quality.

The Nazi experiment in race biology provoked a revulsion against eugenics that effectively ended the movement. Geneticists dismissed eugenics as pseudoscience, both for its exaggeration of the extent to which intelligence and personality were fixed by heredity and for its naïveté about the complex and mysterious ways in which many genes could interact to determine human traits. In 1966, the British geneticist Lionel Penrose observed that "our knowledge of human genes and their action is still so slight that it is presumptuous and foolish to lay down positive principles for human breeding."

Since then, science has learned much more about the human genome, and advances in biotechnology have granted us a say in the genetic makeup of our offspring. Prenatal testing, for example, can warn parents that their unborn child has a genetic condition like Down syndrome or Tay-Sachs disease, presenting them with the often agonizing option of aborting it. The technique of "embryo selection" affords still greater control. Several embryos are created in vitro from the sperm and eggs of the parents; these embryos are then genetically tested, and the one with the best characteristics is implanted in the mother's womb. Both these techniques can be subsumed under "negative" eugenics, because the genes screened against are those associated with diseases or, potentially, with other conditions the parents might regard as undesirable, like low IQ, obesity, same-sex preference, or baldness. But a revival of "positive" eugenics, à la Galton, can also be seen in the use of reproductive technologies involving egg or sperm donation. Advertisements have appeared in Ivy League newspapers offering as much as fifty thousand dollars to egg donors with the right characteristics, like high SAT scores or blue eyes, and global demand for "Nordic" genes has made Denmark the home of the largest sperm bank in the world.

There is a more radical eugenic possibility on the horizon, one beyond anything Galton envisaged. It would involve shaping the heredity of our descendants by tinkering directly with the genetic material in the cells from which they germinate. This technique, called germ-line engineering, has already been used with several species of mammals, most recently employing the new CRISPR (clustered regularly interspaced short palindromic repeats) technology for targeted gene editing. Proponents of germ-line engineering argue that it is only a matter of time before humans can avail themselves of it. The usual justification for germ-line therapy is its potential for eliminating genetic disorders and diseases, not only in the person who develops from the altered genes, but in all of his or her descendants as well. But it also has the potential to be used for "enhancement." If, for example, researchers identified genes linked with intelligence or athletic ability or happiness, germ-line engineering could allow parents the option of eugenically souping up their children in these respects. While the more cautious advocates of germ-line engineering insist that it only be used to correct genetic defects, opponents worry that it will put us on a slippery slope toward eugenic enhancement. After all, if a child fated to have an abnormally low IQ could be "cured" by a germ-line manipulation, then what parent could resist the temptation to add twenty or so points to his normal child's IQ by a similar tweak?

Galtonian eugenics was wrong because it was based on faulty science and carried out by coercion. But Galton's goal, to breed the barbarism out of humanity, was not despicable. The new eugenics, by contrast, is based on a relatively sound (if still largely incomplete) science, and it is not coercive; it might be called "laissez-faire" eugenics, because decisions about the genetic endowment of children would be left up to their parents. Indeed, the only coercive policy that has been contemplated is the one in which the state outlaws these technologies, thereby enforcing the natural genetic order. (Europe, unlike the United States, has already banned germ-line manipulation.)

It is the goal of the new eugenics that is morally cloudy. If its technologies are used to shape the genetic endowment of children according to the desires (and financial means) of their parents, the outcome could be a "GenRich" class of people who are smarter, healthier, and handsomer than the underclass of "Naturals." The ideal of individual enhancement, rather than species uplift, is in stark contrast to the Galtonian vision of eugenics.

"The improvement of our stock seems to me one of the highest objects that we can reasonably attempt," Galton declared in his 1904 address on the aims of eugenics. "We are ignorant of the ultimate destinies of humanity, but feel perfectly sure that it is as noble a work to raise its level . . . as it would be disgraceful to abase it." It might be right to dismiss this (as Martin Brookes does) as a "blathering sermon." But Galton's words possess a certain rectitude when set beside the new eugenicists' talk of a "posthuman" future of designer babies. Galton, at least, had the excuse of historical innocence.

PART III

Mathematics, Pure and Impure

6

A Mathematical Romance

For those who have learned something of higher mathematics, nothing could be more natural than to use the word "beautiful" in connection with it. Mathematical beauty, like the beauty of, say, a late Beethoven quartet, arises from a combination of strangeness and inevitability. Simply defined abstractions disclose hidden quirks and complexities. Seemingly unrelated structures turn out to have mysterious correspondences. Uncanny patterns emerge, and they remain uncanny even after being underwritten by the rigor of logic.

So powerful are these aesthetic impressions that one great mathematician, G. H. Hardy, declared that beauty, not usefulness, is the true justification for mathematics. To Hardy, mathematics was first and foremost a creative art. "The mathematician's patterns, like the painter's or the poet's, must be beautiful," he wrote in his classic 1940 book *A Mathematician's Apology*. "Beauty is the first test: there is no permanent place in the world for ugly mathematics."

And what is the appropriate reaction when one is confronted by mathematical beauty? Pleasure, certainly; awe, perhaps. Thomas Jefferson wrote in his seventy-sixth year that contemplating the truths of mathematics helped him to "beguile the wearisomeness of declining life." To Bertrand Russell—who rather melodramatically claimed, in his autobiography, that it was his desire to know more of mathematics that kept him from committing suicide—the beauty of mathematics was "cold and austere, like that of sculpture . . . sublimely pure, and capable of a stern perfection." For others, mathematical beauty may evoke a distinctly warmer sensation. They might take their cue from Plato's *Symposium*. In that dialogue, Socrates tells the guests assembled at a banquet how a priestess named Diotima initiated him into the mysteries of Eros—the Greek name for desire in all its forms.

One form of Eros is the sexual desire aroused by the physical beauty of a particular beloved person. That, according to Diotima, is the lowest form. With philosophical refinement, however, Eros can be made to ascend toward loftier and loftier objects. The penultimate of these—just short of the Platonic idea of beauty itself—is the perfect and timeless beauty discovered by the mathematical sciences. Such beauty evokes in those able to grasp it a desire to reproduce—not biologically, but intellectually, by begetting additional "gloriously beautiful ideas and theories." For Diotima, and presumably for Plato as well, the fitting response to mathematical beauty is the form of Eros we call love. (In one of those pointless but amusing coincidences, G. H. Hardy tells us near the end of *A Mathematician's Apology* that the Cambridge don who first opened his eyes to the beauty of mathematics was "Professor Love.")

Edward Frenkel, a Russian mathematical prodigy who became a professor at Harvard at twenty-one and who now teaches at Berkeley, is an unabashed Platonist. Eros pervades his winsome 2013 memoir, *Love and Math*, a sort of Platonic love letter to mathematics. As a boy, he was hit by the beauty of mathematics like a *coup de foudre*. When, while still in his teens, he made a new mathematical discovery, it was "like the first kiss." Even when his career hopes seemed blighted by Soviet anti-Semitism, he was sustained by the "passion and joy of doing mathematics."

Frenkel wants everybody to share that passion and joy. Therein lies a challenge. Mathematics is abstract and difficult; its beauties would seem to be inaccessible to most of us. As the German poet Hans Magnus Enzensberger has observed, mathematics is "a blind spot in our culture—alien territory, in which only the elite, the initiated few have managed to entrench themselves." People who are otherwise cultivated will proudly confess their philistinism when it comes to mathematics. The problem is that they have never been introduced to its masterpieces. The mathematics taught in school, and even in college (through, say, introductory calculus), is mostly hundreds or thousands of years old, and much of it involves routine problem solving by tedious calculation.

That bears scant resemblance to what most mathematicians do today. Around the middle of the nineteenth century, a sort of revolution occurred in mathematics: the emphasis shifted from science-bound calculation to the free creation of new structures, new languages. Mathematical proofs, for all their rigorous logic, came to look more like narratives, with plots and subplots, twists and resolutions. It is this kind of mathematics that most people never see. True, it can be daunting. But great works of art, even when difficult, often allow the untutored a glimpse into their beauty. You don't have to know the theory of counterpoint to be moved by a Bach fugue.

Frenkel's own pursuit of the beauty of higher mathematics has led him to play a critical role in the most exciting mathematical drama of the last half century: the Langlands program. Conceived in the 1960s by Robert Langlands, a Canadian mathematician at the Institute for Advanced Study in Princeton (and the inheritor of Einstein's old office there), the Langlands program aims at being a grand unifying theory. For Frenkel, it contains "the source code of all mathematics." Yet it is little known outside the mathematical community. Indeed, most professional mathematicians were unaware of the Langlands program as late as the 1990s, when it figured in the headline-making resolution of Fermat's last theorem. Since then, its scope has expanded beyond pure mathematics to the frontiers of theoretical physics.

Frenkel grew up during the Brezhnev era in an industrial town called Kolomna, about seventy miles outside Moscow. "I hated math when I was at school," he tells us. "What really excited me was physics—especially quantum physics." In his early teens, he avidly read popular physics books that contained titillating references to subatomic particles like hadrons and quarks. Why, he wondered, did the fundamental particles of nature come in such bewildering varieties? Why did they fall into families of certain sizes? It was only when his parents (both industrial engineers) arranged for him to meet with an old friend of theirs, a mathematician, that Frenkel was enlightened. What brought order and logic to the building blocks of matter, the mathematician explained to him, was something called a "symmetry group"—a mathematical beast that Frenkel had never encountered in school.

"This was a moment of epiphany," he recalls, a vision of "an entirely different world."

To a mathematician, a "group" is a set of actions or operations that hang together in a nice way. What is meant by "in a nice way" is spelled out in the four axioms of group theory, which define the algebraic structure of a group. One of the axioms, for example, says that for any action in the group, there is another action in the group that undoes it.

An important kind of group—the kind Frenkel first encountered—is a *symmetry* group. Suppose you have a square card table sitting in the middle of a room. Intuitively, this piece of furniture is symmetrical in certain ways. How can this claim be made more precise? Well, if you rotate the table about its center by exactly 90 degrees, its appearance will be unchanged; no one who was out of the room when the table was rotated will notice any difference upon returning (assuming there are no stains or scratches on its surface). The same is true if you rotate the card table by 180 degrees, or by 270 degrees, or by 360 degrees—the last of which, because it takes the card table in a complete circle, is equivalent to no rotation at all.

These actions constitute the symmetry group of the card table. Since there are only four of them, the group is finite. If the table were circular, by contrast, its symmetry group would be infinite, since any rotation at all—by 1 degree, by 45 degrees, by 132.32578 degrees, or whatever—would leave its appearance unchanged. Groups are thus a way of measuring the symmetry of an object: a circular table, with its infinite symmetry group, is more symmetrical than a square table, whose symmetry group contains just four actions.

But (fortunately) it gets more interesting than that. Groups can capture symmetries that go beyond the merely geometric—like the symmetries hidden in an equation or in a family of subatomic particles. The real power of group theory was first demonstrated in 1832, in a letter that a twenty-year-old Parisian student and political firebrand named Évariste Galois hastily scrawled to a friend late on the night before he was to die in a duel (over the honor of a woman and quite possibly at the hand of a government agent provocateur).

What Galois saw was a truly beautiful way to extend the symmetry concept into the realm of numbers. By his *théorie des groupes*, he was able to resolve a classical problem in algebra that had bedeviled mathematicians for centuries—and in an utterly unexpected way. ("Galois did not *solve* the problem," Frenkel observes. "He *hacked* the problem.") The significance of Galois's discovery far transcended the problem that inspired it. Today, "Galois groups" are ubiquitous in the literature, and the group idea has proved to be perhaps the most versatile in all mathematics, clarifying many a deep mystery. "When in doubt," the great André Weil advised, "look for the group!" That's the *cherchez la femme* of mathematics.

Once smitten, the young Frenkel became obsessed with learning as much of mathematics as he could. ("This is what happens when you fall in love.") When he reached the age of sixteen, it was time to apply to a university. The ideal choice was obvious: Moscow State University, whose department of mechanics and mathematics, nicknamed Mekh-Mat, was one of the great world centers for pure mathematics. But it was 1984, a year before Gorbachev came to power, and the Communist Party still reached into all aspects of Russian life, including university admissions. Frenkel had a Jewish father, and that, apparently, was enough to scupper his chances of getting into Moscow State. (The unofficial rationale for keeping Jews out of physics-related academic areas was that they might pick up nuclear expertise and then immigrate to Israel.) But the appearance of fairness was maintained. He was allowed to sit for the entrance exam, which turned into a sadistic five-hour ordeal out of *Alice's Adventures in Wonderland*. (Interrogator: "What is the definition of a circle?" Frenkel: "A circle is the set of points on the plane equidistant from a given point." Interrogator: "Wrong! It's the set of *all* points on the plane equidistant from a given point.")

Frenkel's consolation prize was a place at the Moscow Institute of Oil and Gas (cynically nicknamed Kerosinka), which had become a haven for Jewish students. But such was his craving for pure mathematics, he tells us, that he would scale a twenty-foot fence at the heavily guarded Mekh-Mat to get into the seminars there. Soon his extraordinary ability was recognized by a leading figure in Moscow mathematics, and he was put to work on an unsolved problem, which engrossed him for weeks to the point of insomnia. "And then, suddenly, I had it," he recalls. "For the first time in my life, I had in my possession something that no one else in the world had." The problem he had solved concerned yet another species of abstract group, called braid groups because they arise from systems of entwined curves that look quite literally like braided hair.

Despite this and other breakthroughs that Frenkel made while still in his late teens, his academic prospects as a quasi Jew were dim. But his talent had come to the attention of mathematicians abroad. In 1989, the mail brought an unexpected letter from Derek Bok, the president of Harvard. The letter addressed Frenkel as "Doctor" (though he did not yet possess so much as a bachelor's degree) and invited him to come to Harvard as a prize fellow. "I had heard about Harvard University before," Frenkel recalls, "though I must admit I did not quite realize at the time its significance in the academic world." At the age of twenty-one, he would be a visiting professor of mathematics at Harvard, with no formal obligations except to give occasional lectures about his work. And to his equal amazement, he received a Soviet exit visa in a month, becoming one of the first in what would be an exodus of Jewish mathematicians in the age of perestroika.

Frenkel's adjustment to American life was reasonably smooth. He marveled at the "abundance of capitalism" in the aisles of a Boston supermarket; he bought "the hippest jeans and a Sony Walkman"; he struggled to learn the ironic nuances of English by devotedly watching the David Letterman show on TV every night. Most important, he met another Russian Jewish émigré at Harvard who introduced him to the Langlands program.

As with Galois theory, the Langlands program had its origins in a letter. It was written in 1967 by Robert Langlands (then in his early thirties) to one of his colleagues at the Institute for Advanced Study, André Weil. In his letter, Langlands proposed the possibility of a deep analogy between two theories that seemed to lie at opposite ends of the mathematical cosmos: the theory of Galois groups, which concerns symmetries in the realm of numbers, and "harmonic analysis," which concerns how complicated waves (for example, the sound of a symphony) are built up from simple harmonics (for example, the individual instruments). Certain structures in the harmonic world, called automorphic forms, somehow "knew" about mysterious patterns in the world of numbers. Thus it might be possible to use the methods of one world to reveal hidden harmonies in the other—so Langlands conjectured. If Weil did not find the intuitions in the letter persuasive, Langlands added, "I am sure you have a waste basket handy."

But Weil, a magisterial figure in twentieth-century mathematics (he died in 1998 at the age of ninety-two), was a receptive audience. In a letter he had written in 1940 to his sister, Simone Weil, he had described in vivid terms the importance of analogy in mathematics. Alluding to the Bhagavad Gita (he was also a Sanskrit scholar), André explained to Simone that just as the Hindu deity Vishnu had ten different avatars, a seemingly simple mathematical equation could manifest itself in dramatically different abstract structures. The subtle analogies between such structures were like "illicit liaisons," he wrote; "nothing gives more pleasure to the connoisseur." As it happens, Weil was writing to his sister from prison in France, where he had been temporarily confined for desertion from the army (after nearly being executed as a spy in Finland).

The Langlands program is a scheme of conjectures that would turn such hypothetical analogies into sturdy logical bridges, linking up diverse mathematical islands across the surrounding sea of ignorance. Or it can be seen as a Rosetta stone that would allow the mathematical tribes on these various islands—number theorists, topologists, algebraic geometers—to talk to one another and pool their conceptual resources. The Langlands conjectures are largely unproved so far. (An exception is the Taniyama-Shimura conjecture, framed in the 1950s by a pair of Japanese mathematicians and proved in the 1990s by the Englishman Andrew Wiles, who thereby established the truth of Fermat's last theorem.) Are these mysterious conjectures even true? There is an almost Platonic confidence among mathematicians that they must be. As Ian Stewart has remarked, the Langlands program is "the sort of mathematics that ought to be true because it was so beautiful." The unity it could bring to higher mathematics could usher in a new golden age in which we may finally discover, as Frenkel puts it, "what mathematics is really about."

Since Frenkel had no graduate degree, he had to undergo a temporary "demotion" from Harvard professor to graduate student while he wrote a Ph.D. thesis, which he wrapped up in a single year. (At his 1991 graduation, he was pleased to be personally congratulated by one of the honorary-degree recipients that year, Eduard Shevardnadze, an architect of perestroika.) In his thesis, Frenkel proved a theorem that helped open a new chapter in the Langlands program, extending it from the realm of numbers into the geometric realm of curved surfaces, like the surface of a ball or a donut. (These are called Riemann surfaces, after the nineteenth-century mathematician Bernhard Riemann.)

The pursuit of the Langlands program has involved twisting, even shattering, many familiar mathematical ideas—ideas as basic as the counting numbers. Consider the number 3. It's boring; it has no internal structure. But suppose you replace the number 3 with a "vector space" of three dimensions—that is, a space in which each point represents a trio of numbers, with its own rules for addition and multiplication. Now you've got something interesting: a structure with more symmetries than a Greek temple. "In modern math, we create a new world in which numbers come alive as vector spaces," Frenkel writes. And other basic concepts are enriched too. The "functions" that you might have run into in high school mathematics—as in $y = f(x)$—are transformed into exotic creatures called sheaves. (The man most responsible for this reinvention of the language of mathematics was Alexander Grothendieck, generally considered the greatest mathematician of the latter half of the twentieth century.)

The next move was to extend the Langlands program beyond the borders of mathematics itself. In the 1970s, it had been noticed that one of its key ingredients—the "Langlands dual group"—also crops up in quantum physics. This came as a surprise. Could the same patterns that can be dimly glimpsed in the worlds of number and geometry also have counterparts in the theory that describes the basic forces of nature? Frenkel was struck by the potential link between quantum physics and the Langlands program and set about to investigate it—aided by a multimillion-dollar grant that he and some colleagues received in 2004 from the Department of Defense, the largest grant to date for research in pure mathematics. (In addition to being clean and gentle, pure mathematics is cheap: all its practitioners need is chalk and a little travel money. It is also open and transparent, because there are no inventions to patent.)

This brought Frenkel into a collaboration with Edward Witten, widely regarded as the greatest living mathematical physicist (and, like Langlands himself, a member of the Institute for Advanced Study in Princeton). Witten is a virtuoso of string theory, an ongoing effort by physicists to unite all the forces of nature, including gravity, in one neat mathematical package. He awed Frenkel with his "unbreakable logic" and his "great taste." It was Witten who saw how the "branes" (short for "membranes") postulated by string theorists might be analogous to the "sheaves" invented by mathematicians. Thus opened a rich dialogue between the Langlands program, which aims to unify mathematics, and string theory, which aims to unify physics. Although optimism about string theory has faded somewhat with its failure (thus far) to deliver an effective description of our universe, the Langlands connection has yielded deep insights into the workings of particle physics.

This is not the first time that mathematical concepts studied for their pure beauty have later turned out to illumine the physical world. "How can it be," Einstein asked in wonderment, "that mathematics, being after all a product of human thought independent of experience, is so admirably appropriate to the objects of reality?" Frenkel's take on this is very different from Einstein's. For Frenkel, mathematical structures are among the "objects of reality"; they are every bit as real as anything in the physical or mental world. Moreover, they are not the product of human thought; rather, they exist timelessly, in a Platonic realm of their own, waiting to be discovered by mathematicians. The conviction that mathematics has a reality that transcends the human mind is not uncommon among its practitioners, especially great ones like Frenkel and Langlands, Sir Roger Penrose and Kurt Gödel. It derives from the way that strange patterns and correspondences unexpectedly emerge, hinting at something hidden and mysterious. Who put those patterns there? They certainly don't seem to be of our making.

The problem with this Platonist view of mathematics—one that Frenkel, going on in a *misterioso* vein, never quite recognizes as such— is that it makes mathematical knowledge a miracle. If the objects of mathematics exist apart from us, living in a Platonic heaven that transcends the physical world of space and time, then how does the human mind "get in touch" with them and learn about their properties and relations? Do mathematicians have ESP? The trouble with Platonism, as the philosopher Hilary Putnam has observed, "is that it seems flatly incompatible with the simple fact that we think with our brains, and not with immaterial souls."

Perhaps Frenkel should be allowed his Platonic fantasy. After all, every lover harbors romantic delusions about his beloved. In 2009, while Frenkel was in Paris as the occupant of the *Chaire d'Excellence* of the Fondation Sciences Mathématiques, he decided to make a short film expressing his passion for mathematics. Inspired by Yukio Mishima's *Rite of Love and Death*, he titled it *Rites of Love and Math*. In this silent Noh-style allegory, Frenkel plays a mathematician who creates a formula of love. To keep the formula from falling into evil hands, he hides it away from the world by tattooing it with a bamboo stick on the body of the woman he loves and then prepares to sacrifice himself for its protection.

Upon the premiere of *Rites of Love and Math* in Paris in 2010, *Le Monde* called it "a stunning short film" that "offers an unusual romantic vision of mathematicians." The "formula of love" used in the film was one that Frenkel himself discovered (in the course of investigating the mathematical underpinnings of quantum field theory). It is beautiful yet forbidding. The only numbers in it are 0, 1, and ∞. Isn't love like that?

7

The Avatars of Higher Mathematics

"The science of pure mathematics . . . may claim to be the most original creation of the human spirit." So declared the philosopher (and lapsed mathematician) Alfred North Whitehead. Strange, then, that the practitioners of this "science" still feel the need to justify their vocation—not to mention the funding that the rest of society grants them to pursue it. Note that Whitehead said "pure" mathematics. He wasn't talking about the "applied" variety: the kind that is cultivated for its usefulness to the empirical sciences or for commercial purposes. (The latter is sometimes disparagingly referred to as industrial mathematics.) "Pure" mathematics is indifferent to such concerns. Its deepest problems arise out of its own inward-looking mysteries.

From time to time, of course, research in pure mathematics does turn out to have applications. The theoretical goose lays a golden egg. It was to this potential for unexpectedly useful by-products that Abraham Flexner, the founder of the Institute for Advanced Study in Princeton, called attention in a 1939 article in *Harper's Magazine* titled "The Usefulness of Useless Knowledge." But the "Golden Goose argument" (as the Harvard historian Steven Shapin has dubbed it) is not one that much appeals to pure mathematicians. The British mathematician G. H. Hardy, for one, was positively contemptuous of the idea that "real" mathematics should be expected to have any practical importance.

In his 1940 book *A Mathematician's Apology*—justly hailed by David Foster Wallace as "the most lucid English prose work ever on math"—Hardy argued that the point of mathematics was the same as the point of art: the creation of intrinsic beauty. He reveled in what he presumed to be the utter uselessness of his own specialty, the theory of numbers. No doubt Hardy, who died in 1947, would be distressed to learn that his "pure" number theory has been pressed into impure service as the basis for public-key cryptography, which allows customers to send encrypted credit card information to an online store without any exchange of secret cryptographic keys, thus making trillions of dollars' worth of e-commerce possible, and that his work in a branch of mathematics called functional analysis proved fundamental to the notorious Black-Scholes equation, used on Wall Street to price financial derivatives.

The irony of pure mathematics begetting crass commercialism is not lost on Michael Harris, whose memoir, *Mathematics Without Apologies*, irreverently echoes Hardy's classic title. Harris is a distinguished middle-aged American mathematician who works in the gloriously pure stratosphere where algebra, geometry, and number theory meet. "The guiding problem for the first part of my career," he writes, was "the Conjecture of Birch and Swinnerton-Dyer," which "concerns the simplest class of polynomial equations—elliptic curves—for which there is no simple way to decide whether the number of solutions is finite or infinite." (Though elementary in appearance, elliptic curves turn out to have a deep structure that makes them endlessly interesting.) He has spent much of his research career in Paris, and it shows; his memoir is full of Gallic intellectual playfulness, plus references to figures like Pierre Bourdieu, Issey Miyake, and Catherine Millet ("the sexual Stakhanovite") and mention of the endless round of Parisian champagne receptions where "mathematical notes are compared for the first glass or two, after which conversation reverts to university politics and gossip." It is rambling, sardonic (the term "fuck-you money" appears in the index), and witty. It contains fascinating literary digressions, such as an analysis of the occult mathematical structure of Thomas Pynchon's novels, and lovely little interludes on elementary math, inspired by Harris's gallant attempt to explain number theory to a British actress at a Manhattan dinner party.

Starting with the simple definition of a prime number, he builds, bit by bit, to an explanation of the aforementioned Birch and Swinnerton-

Dyer conjecture—which, at a press conference given in Paris in the year 2000 by an international group of leading mathematicians, was declared one of seven "Millennium Prize Problems," whose resolution would be rewarded by a million-dollar prize. Harris takes an intimate look at the deepest developments in contemporary mathematics, especially the visionary work of Alexander Grothendieck. And he makes much of what he calls the "pathos" of the mathematician's calling.

Harris is rudely skeptical of the usual justifications for pure mathematics: that it is beautiful, true, or even much good, at least in the utilitarian "Golden Goose" sense. "It is not only dishonest but also self-defeating to pretend that research in pure mathematics is motivated by potential applications," he observes. He notes that public-key cryptography, by making the world safe for Amazon, has destroyed the corner bookstore (in America, not France, where online retailers are prohibited by law from offering free shipping on discounted books). And he displays an Olympian scorn for the sudden popularity of "finance mathematics," which offers a path to derivative-fueled wealth on Wall Street: "A colleague boasted that Columbia's mathematical finance program was underwriting the lavish daily spreads of fresh fruit, cheese, and chocolate brownies, when other departments, including mine in Paris, were lucky to offer a few teabags and a handful of cookies to calorie-starved graduate students." Even at France's elite École Polytechnique, 70 percent of the mathematics students today aspire to a career in finance.

Nor is Harris impressed with the claim, voiced by Hardy and so many others, that pure mathematics is justified by its beauty. When mathematicians talk about beauty, he tells us, what they really mean is pleasure. "Outside this relaxed field, it's considered poor form to admit that we are motivated by pleasure," Harris writes. "Aesthetics is a way of reconciling this motivation with the 'lofty habit of mind.'"

Why should society pay for a small group of people to exercise their creative powers on something they enjoy? "If a government minister asked me that question," Harris replies, "I could claim that mathematicians, like other academics, are needed in the universities to teach a specific population of students the skills needed for the development of a technological society and to keep a somewhat broader population of students occupied with courses that serve to crush the dreams of superfluous applicants to particularly desirable professions (as freshman calculus used to be a formal requirement to enter medical school in the United States)." Although physicians don't really need calculus, Harris at least concedes that engineers, economists, and inventory managers couldn't get by without a fair amount of math, even if it is trivial math by his lights.

Finally, there is the presumed value of mathematical truth. Since the ancient Greeks, mathematics has been taken as a paradigm of knowledge: certain, timeless, necessary. But knowledge of what? Do the truths discovered by mathematics describe an eternal and otherworldly realm of objects—perfect circles and so forth—that exist quite independently of the mathematicians who contemplate them? Or are mathematical objects actually human constructions, existing only in our minds? Or, more radically still, could it be that pure mathematics doesn't really describe any objects at all, that it is just an elaborate game of formal symbols played with pencil and paper?

The question of what mathematics is really about is one that continues to vex philosophers, but it does not much worry Harris. Philosophers who concern themselves with the problems of mathematical existence and truth, he claims, typically pay little attention to what mathematicians actually do. He invidiously contrasts what he calls "philosophy of Mathematics" (with a capital M)—"a purely hypothetical subject invented by philosophers"—with "philosophy of (small-m) mathematics," which takes as its starting point not a priori questions about epistemology and ontology but rather the activity of working mathematicians.

Here, Harris is being a little unfair. He fails to remark that the standard competing positions in the philosophy of mathematics were originally staked out not by philosophers but by mathematicians—indeed, some of the greatest of the last century. It was David Hilbert—a "supergiant," in Harris's estimation—who originated "formalism," which views higher mathematics as a game played with formal symbols. Henri Poincaré (another "supergiant"), Hermann Weyl, and L.E.J. Brouwer were behind "intuitionism," according to which numbers and other mathematical objects are mind-dependent constructions. Bertrand Russell and Alfred North Whitehead took the position known as "logicism," endeavoring to show in their massive *Principia Mathematica* that mathematics was really logic in disguise. And

"Platonism"—the idea that mathematics describes a perfect and eternal realm of mind-independent objects, like Plato's world of forms—was championed by Kurt Gödel.

All of these mathematical figures were passionately engaged in what Harris slights as philosophy of Mathematics-with-a-capital-M. The debate among them and their partisans was fierce in the 1920s, often spilling over into personal animus. And no wonder: mathematics at the time was undergoing a "crisis" that had resulted from a series of confidence-shaking developments, like the emergence of non-Euclidean geometries and the discovery of paradoxes in set theory. If the old ideal of certainty was to be salvaged, it was felt, mathematics had to be put on a new and secure foundation. At issue was the very way mathematics would be practiced: what types of proof would be accepted as valid, what uses of infinity would be permitted.

For reasons both technical and philosophical, none of the competing foundational programs of the early twentieth century proved satisfactory. (Gödel's "incompleteness theorems," in particular, created insuperable problems both for Hilbert's formalism and for Russell and Whitehead's logicism: the incompleteness theorems showed—roughly speaking—that the rules of Hilbert's mathematical "game" could never be proved consistent and that a logical system like that of Russell and Whitehead could never capture all mathematical truths.) The issues of mathematical existence and truth remain unresolved, and philosophers have continued to grapple with them, if inconclusively—witness the frank title that Hilary Putnam gave to a 1979 paper: "Philosophy of Mathematics: Why Nothing Works."

To Harris, this looks a bit *vieux jeu*. The sense of crisis in the profession, so acute less than a century ago, has receded; the old difficulties have been patched up or papered over. If you ask a contemporary mathematician to declare a philosophical party affiliation, the joke goes, you'll hear "Platonist" on weekdays and "formalist" on Sundays: that is, when they're working at mathematics, mathematicians regard it as being about a mind-independent reality; but when they're in a reflective mood, many claim to believe that it's just a meaningless game played with formal symbols.

Today, paradigm shifts in mathematics have less to do with "crisis" and more to do with finding superior methods. It used to be thought, for example, that all mathematics could be constructed out of sets. Starting with the simple idea of one thing being a member of another, set theory shows how structures of seemingly limitless complexity—number systems, geometric spaces, a never-ending hierarchy of infinities—can be built up out of the most modest materials. The number 0, for example, can be defined as the "empty set": that is, the set that has no members at all. The number 1 can then be defined as the set that contains one element—0 and nothing else. Two, in turn, can be defined as the set that contains 0 and 1—and so on, with the set for each subsequent number containing the sets for all the previous numbers. Numbers, instead of being taken as basic, can thus be viewed as pure sets of increasingly intricate structure.

In the 1930s, a cabal of brilliant young Paris mathematicians, including André Weil, resolved to make the house of mathematics more secure by rebuilding it on the logical foundation of set theory. The project, under the collective nom de guerre Bourbaki, went on for decades, resulting in one fat treatise after another. Among its consequences, crazily enough, was the advent of the "new math" education reforms back in the 1960s, which so befuddled American schoolchildren and their parents by replacing intuitive talk of numbers with the alien jargon of sets.

Physicists talk about finding the "theory of everything"; well, set theory is so sweeping in its generality that it might appear to be "the theory of theories of everything." It certainly appeared that way to the members of Bourbaki. Yet a few decades after their program got under way, the extraordinary Alexander Grothendieck came into their midst and transcended it. In doing so, he created a new style of pure mathematics that proved as fruitful as it was dizzyingly abstract. Long before his death in 2014 at the age of eighty-six in a remote hamlet in the Pyrenees, Grothendieck had come to be regarded as the greatest mathematician of the last half century. As Harris observes, he likely qualifies as the "most romantic" too: "His life story begs for fictional treatment."

The raw facts are astounding enough. Alexander Grothendieck was born in Berlin in 1928 to parents who were active anarchists. His father, a Russian Jew, took part in the 1905 uprising against the tsarist regime and the 1917 revolution. Grothendieck's father escaped imprisonment under the Bolsheviks; clashed with Nazi thugs on the streets of Berlin; fought on the Republican side in the Spanish Civil War (as did Grothendieck's mother); and was deported, after the fall of France, from Paris to Auschwitz to be murdered.

Grothendieck's mother, a gentile from Hamburg, raised him in the South of France. There the boy showed talent both for numbers and for boxing. After the war, he made his way to Paris to study mathematics under the great Henri Cartan. Following early teaching stints in São Paulo and Kansas, and at Harvard, Grothendieck was invited in 1958 to join the Institut des Hautes Études Scientifiques, which had just been founded by a private businessman outside Paris in the woods of Bois-Marie. There Grothendieck spent the next dozen years astonishing his elite colleagues and younger disciples as he re-created the landscape of higher mathematics.

Grothendieck was a physically imposing man, shaven-headed and handsome, as charismatic as he was austere. His ruthless minimalism extended to his contempt for money and his monkish wardrobe. A staunch pacifist and antimilitarist, he refused to go to Moscow in 1966 (where the International Congress of Mathematicians was being held) to accept the Fields Medal, the highest honor in mathematics. He did, however, make a trip the next year to North Vietnam, where he lectured on pure mathematics in the jungle to students who had been evacuated from Hanoi to escape the American bombing. He remained (by choice) stateless most of his life, had three children by a woman he married and two more out of wedlock, founded the radical ecology group Survivre et Vivre, and once got arrested for knocking down a couple of gendarmes at a political demonstration in Avignon.

Owing to his unyielding and sometimes paranoiac sense of integrity, Grothendieck ended up alienating himself from the French mathematical establishment. In the early 1990s, he vanished into the Pyrenees—where, it was reported by the handful of admirers who managed to track him down, he spent his remaining years subsisting on dandelion soup and meditating on how a malign metaphysical force was destroying the divine harmony of the world, possibly by slightly altering the speed of light. The local villagers were said to look after him.

Grothendieck's vision of mathematics led him to develop a new language—it might even be called an ideology—in which hitherto unimaginable ideas could be expressed. He was the first to champion the principle that knowing a mathematical object means knowing its relations to all other objects of the same kind. In other words, if you want to know the real nature of a mathematical object, don't look inside it but see how it plays with its peers.

Such a peer group of mathematical objects is called, in a deliberate nod to Aristotle and Kant, a category. One category might consist of abstract surfaces. These surfaces play together, in the sense that there are natural ways of going back and forth between them that respect their general form. For example, if two surfaces have the same number of holes—like a donut and a coffee mug—one surface can, mathematically, be smoothly transformed into the other.

Another category might consist of all the different algebraic systems that have an operation akin to multiplication; these algebras too play together, in the sense that there are natural ways of going back and forth between them that respect their common multiplicative structure. Such structure-preserving back-and-forth relations among objects in the same category are called morphisms, or sometimes—to stress their abstract nature—arrows. They determine the overall shape of the play within a category.

And here's where it gets interesting: the play in one category—say, the category of surfaces—might be subtly mimicked by the play in another—say, the category of algebras. The two categories themselves can be seen to play together: there is a natural way of going back and forth between them, called a functor. Armed with such a functor, one can reason quite generally about both categories, without getting bogged down in the particular details of each. It might also be noticed that because categories play with each other, they themselves form a category: the category of categories.

Category theory was invented in the 1940s by Saunders Mac Lane of the University of Chicago and Samuel Eilenberg of Columbia. At first it was regarded dubiously by many mathematicians, earning the nickname "abstract nonsense." How could such a rarefied approach to mathematics, in which nearly all its classical content seemed to be drained away, result in anything but sterility? Yet Grothendieck made it sing. Between 1958 and 1970, he used category theory to create novel structures of unexampled richness. Since then, the heady abstractions of category theory have also become useful in theoretical physics, computer science, logic, and philosophy. The French philosopher Alain Badiou has been drawing on category theory since the 1980s— and in a mathematically informed way—to explore ideas of being and transcendence.

The project undertaken by Grothendieck was one that began with Descartes: the unification of geometry and algebra. These have been likened to the yin and yang of mathematics: geometry is space, algebra is time; geometry is like painting, algebra is like music; and so on. Less fancifully, geometry is about form, algebra is about structure—in particular, the structure that lurks within equations. And as Descartes showed with his invention of "Cartesian coordinates," equations can describe forms: $x^2 + y^2 = 1$, for example, describes a circle of radius 1. So algebra and geometry turn out to be intimately related, exchanging what André Weil called "subtle caresses."

In the 1940s, thanks to Weil's insight, it became apparent that the dialectic between geometry and algebra was the key to resolving some of the most stubbornly enduring mysteries in mathematics. And it was Grothendieck's labor that raised this dialectic to such a pitch of abstraction—one that was said to leave even the great Weil daunted— that a new understanding of these mysteries emerged. Grothendieck laid the groundwork for many of the greatest mathematical advances in recent decades, including the 1994 proof of Fermat's last theorem—a magnificent intellectual achievement of zero practical or commercial interest.

Grothendieck transformed modern mathematics. However, much of the credit for this transformation should go to a lesser-known forerunner of his, Emmy Noether. It was Noether, born in Bavaria in 1882, who largely created the abstract approach that inspired category theory.[*] Yet as a woman in a male academic world, she was barred from holding a professorship in Göttingen, and the classicists and historians on the faculty even tried to block her from giving unpaid lectures—leading David Hilbert, the dean of German mathematics, to comment, "I see no reason why her sex should be an impediment to

---

[*]And this was only one of her towering accomplishments—see "Emmy Noether's Beautiful Theorem" in this volume, p. 274.

her appointment. After all, we are a university, not a bathhouse."
Noether, who was Jewish, fled to the United States when the Nazis took
power, teaching at Bryn Mawr until her death from a sudden infection
in 1935.

The intellectual habit of grappling with a problem by ascending to
higher and higher levels of generality came naturally to Emmy Noether,
and it was shared by Grothendieck, who said that he liked to solve a
problem not by the "hammer-and-chisel method" but by letting a sea
of abstraction rise to "submerge and dissolve" it. In his vision, the familiar things dealt with by mathematicians, like equations, functions,
and even geometric points, were reborn as vastly more complex and
versatile structures. The old things turned out to be mere shadows—
or, as Grothendieck preferred to call them, "avatars"—of the new. (An
avatar is originally an earthly manifestation of a Hindu god; perhaps
because of the influence of André Weil, who was also an expert in
Sanskrit, many French mathematicians have a terminological predi-
lection for Hindu metaphysics.)

Nor is this a one-off process. Each new abstraction is eventually
revealed to be but an avatar of a still-higher abstraction. As Michael
Harris puts it, "The available concepts are interpreted as the avatars of
the inaccessible concepts we are trying to grasp." With the grasping of
these new concepts, mathematics ascends a kind of "ladder" of increasing abstraction. And it is this, Harris insists, that philosophers
should be paying attention to: "If you were to ask for a single characteristic of contemporary mathematics that cries out for philosophical
analysis, I would advise you to practice climbing the categorical and
avatar ladders in search of meaning, rather than searching for solid
Foundations."

And what lies at the top of this ladder? Perhaps, Harris suggests
with playful seriousness, there is "One Big Theorem" from which all of
mathematics ultimately flows—"something on the order of *samsara* =
nirvana." But since there are infinitely many rungs to climb, it is unat-
tainable.

Here, then, is the pathos of mathematics. Unlike theoretical physics, which can aspire to a "final theory" that would account for all the
forces and particles in the universe, pure mathematics must concede
the futility of its own quest for ultimate truth. As Harris observes,

"Every veil lifted only reveals another veil." The mathematician is doomed to what André Weil called an endless cycle of "knowledge and indifference."

But it could be worse. Thanks to Gödel's second incompleteness theorem—the one that says, roughly, that mathematics can never prove its own consistency—mathematicians can't be fully confident that the axioms underlying their enterprise do not harbor an as-yet-undiscovered logical contradiction. This possibility is "extremely unsettling for any rational mind," declared the Russian-born mathematician (and Fields medalist) Vladimir Voevodsky, in a speech on the occasion of the eightieth anniversary of the Institute for Advanced Study. Indeed, the discovery of such an inconsistency would be fatal to pure mathematics, at least as we know it today. The distinction between truth and falsehood would be breached, the ladder of avatars would come crashing down, and the One Big Theorem would take a truly terrible form: $0 = 1$.

Yet, oddly enough, e-commerce and financial derivatives would be left untouched.

8

Benoit Mandelbrot and the Discovery of Fractals

Benoit Mandelbrot, the brilliant Polish-French-American mathematician who died in 2010, had a poet's taste for complexity and strangeness. His genius for noticing deep links among far-flung phenomena led him to create a new branch of geometry, one that has deepened our understanding of both natural forms and patterns of human behavior. The key to it is a simple yet elusive idea, that of self-similarity.

To see what self-similarity means, consider a homely example: the cauliflower. Take a head of this vegetable and observe its form—the way it is composed of florets. Pull off one of those florets. What does it look like? It looks like a little head of cauliflower, with its own sub-florets. Now pull off one of those sub-florets. What does that look like? A still-tinier cauliflower. If you continue this process—and you may soon need a magnifying glass—you'll find that the smaller and smaller pieces all resemble the head you started with. The cauliflower is thus said to be self-similar. Each of its parts echoes the whole.

Other self-similar phenomena, each with its distinctive form, include clouds, coastlines, bolts of lightning, clusters of galaxies, the network of blood vessels in our bodies, and, quite possibly, the pattern of ups and downs in financial markets. The closer you look at a coastline, the more you find it is jagged, not smooth, and each jagged segment contains smaller, similarly jagged segments that can be described by Mandelbrot's methods. Because of the essential roughness of self-similar forms, classical mathematics is ill-equipped to deal with them. Its methods, from the Greeks on down to the last century, have been better suited to smooth forms, like circles. (Note that a circle is not self-similar: if you cut it up into smaller and smaller segments, those segments become nearly straight.)

Only in the last few decades has a mathematics of roughness emerged, one that can get a grip on self-similarity and kindred matters like turbulence, noise, clustering, and chaos. And Mandelbrot was the prime mover behind it. He had a peripatetic career, but he spent much of it as a researcher for IBM in upstate New York. In the late 1970s, he became famous for popularizing the idea of self-similarity and for coining the word "fractal" (from the Latin *fractus*, meaning "broken") to designate self-similar forms. In 1980, he discovered the "Mandelbrot set," whose shape—it looks a bit like a warty snowman or beetle—came to represent the newly fashionable science of chaos. What is perhaps less well known about Mandelbrot is the subversive work he did in economics. The financial models he created, based on his fractal ideas, implied that stock and currency markets were far riskier than the reigning consensus in business schools and investment banks supposed and that wild gyrations—like the 777-point plunge in the Dow on September 29, 2008—were inevitable.

I was familiar with these aspects of Mandelbrot's career before I read his posthumous 2012 memoir, *The Fractalist: Memoir of a Scientific Maverick*, a draft of which he completed shortly before his death at the age of eighty-five. I knew of his reputation as a "maverick" and "troublemaker"—labels that, despite his years with IBM, seemed well merited. What I wasn't prepared for was the dazzling range of people he intersected with in the course of his career. Consider this partial listing of the figures that crop up in his memoir: Margaret Mead, Valéry Giscard d'Estaing, Claude Lévi-Strauss, Noam Chomsky, Robert Oppenheimer, Jean Piaget, Fernand Braudel, Claudio Abbado, Roman Jakobson, George Shultz, György Ligeti, Stephen Jay Gould, Philip Johnson, and the empress of Japan.

Nor did I realize that Mandelbrot's casually anarchic ways at IBM were at least partly responsible for the advent of that bane of modern life, the computer password. What struck me most, though, was the singularity of Mandelbrot's intuition. Time and again, he found simplicity and even beauty where others saw irredeemable messiness. His secret? A penchant for playing with pictures, a reliance on visual insight: "When I seek, I look, look, look . . ."

Mandelbrot was born in 1924 into a Jewish family that lived in Warsaw. Neither of his parents was mathematical. His father sold ladies' hosiery, and his mother was a dentist—adept, thanks to her "strong right hand and powerful biceps," at pulling teeth. His uncle Szolem, however, was a mathematician of international rank who trained in Paris and became a professor at the Collège de France. "No one would influence my scientific life as much as Szolem," Mandelbrot tells us, though the nature of his uncle's influence would turn out to be rather peculiar.

Describing his Warsaw childhood, he vividly recalls, for example, the manure-like stench given off by one of his mother's dental patients, who defrayed the cost of repairing a mouthful of rotten teeth by bringing the family fresh meat from the slaughterhouse where the patient worked. With the Depression, his father's business collapsed, and eventually the family left Poland for Paris, traveling across Nazi Germany in a padlocked train. "Of the people we knew, we alone moved to France and survived," Mandelbrot writes, adding that many of their neighbors in the Warsaw ghetto "had been detained by their precious china, or inability to sell their Bösendorfer concert grand piano."

Paris enchanted the young Mandelbrot. His family set up housekeeping in a cold-water flat in the then-slummy neighborhood of Belleville, near the Buttes Chaumont, but the boy avidly explored the city at large—the Louvre, the old science museum on the rue St.-Martin, the Latin Quarter. He quickly mastered the new language. One day his father lugged back to the flat "an obsolete multivolume *Larousse Encyclopédie*, together with decades of bound volumes of its updates. In no time, I read them from cover to cover." Although proper French was spoken in the lycée he attended, he picked up in his Belleville neighborhood a sort of cockney Parisian, in which *marrant* (funny) sounded like *marron* (brown). As a result, he says, "my spoken French never quite stabilized, and I keep an accent that varies in time and cannot easily be traced."

In school, Mandelbrot distinguished himself as *un crack*—slang for "high achiever"—and even *un taupin*: "linguistically," he tells us, "an extreme form of the American 'nerd'" (the word derives from the French word *taupe*, meaning "mole"). What gave him an edge was his ability to "geometrize" a problem. Instead of shuffling formulas like his fellow students, he used his prodigious visual memory to see how a complicated equation might harbor a simple shape in disguise. In a nationwide competitive exam, he tells us, he was the only student in France who managed to solve one especially fiendish problem. "How did you manage?" asked his incredulous teacher, a certain Monsieur Pons. "No human could resolve that triple integral in the time allowed!" Mandelbrot informed his teacher that he simply changed the coordinates in which the problem was stated so its geometric essence, that of a sphere, was revealed—whereupon M. Pons walked away muttering, "But of course, of course, of course!"

Where did Mandelbrot's geometric "inner voice" come from? He suggests it may have something to do with his childhood fascination with chess and maps. He also credits the "outdated" math books he happened to get his hands on as a teenage émigré, which had more pictures than those used in school then (and today). "Learning mathematics from such books made me intimately familiar with a large zoo, collected over centuries, of very specialized shapes of every kind," he tells us. "I could recognize them instantly, even when they were dressed up in an analytic garb that was 'foreign' to me and, I thought, to their basic nature."

Mandelbrot was fourteen when World War II broke out. With the fall of Paris, he and his family sought refuge in Vichy France, where, as Jews of foreign origin, they lived in constant fear of denunciation and soon had to split up. Using an assumed name and furnished with fake papers, Mandelbrot pretended to be an apprentice toolmaker in a hardscrabble village in Limousin (where a trace of the rural accent was added to his mix of slum Parisian and correct French). After a close brush with arrest, he made his way to Lyon, where, under the nose of Klaus Barbie, he refined his geometric gift with the help of an inspired teacher at the local lycée.

During this time, Mandelbrot conceived what he calls his "Keplerian quest." Three centuries earlier, Johannes Kepler had made sense of the seemingly irregular motions of the planets by a single geometric insight: he posited that their orbits, instead of being circular as had been supposed since ancient times, took the form of an ellipse. As a teenager, Mandelbrot "came to worship" Kepler's achievement and aspired to do something similar—to impose order on an inchoate area of science through a bold geometric stroke.

It was in postwar Paris that Mandelbrot began this quest in earnest. Uncle Szolem urged him to attend the École Normale Supérieure, France's most rarefied institution of higher learning, where Mandelbrot had earned entry at the age of twenty (one of only twenty Frenchmen to do so). But the aridly abstract style of mathematics practiced there was uncongenial to him. At the time, the École Normale— *dite normale, prétendue supérieure*, says the wag—was dominated in mathematics by a semisecret cabal called Bourbaki. (The name Bourbaki was jocularly taken from a hapless nineteenth-century French general who once tried to shoot himself in the head but missed.) Its leader was André Weil, one of the supreme mathematicians of the twentieth century (and the brother of Simone Weil).

The aim of Bourbaki was to purify mathematics, to rebuild it on perfectly logical foundations untainted by physical or geometric intuition. Mandelbrot found the Bourbaki cult, and Weil in particular, "positively repellent." The Bourbakistes seemed to cut off mathematics from natural science, to make it into a sort of logical theology. They regarded geometry, so integral to Mandelbrot's Keplerian dream, as a dead branch of mathematics, fit for children at best. So, on his second day at the École Normale, Mandelbrot resigned. His uncle was disgusted by his decision, but this only fortified his resolve. Whereas Szolem was a "prudent conformist who promptly joined the soon-to-be-powerful Bourbaki," Mandelbrot saw himself—somewhat megalomaniacally, he concedes—as a "dissenter" who would overturn its orthodoxy.

Groping his way toward this goal, Mandelbrot enrolled in another of France's *grandes écoles*, the École Polytechnique, where the nation's engineering elite was trained. Then located in the Latin Quarter, behind a majestic gate at 5, rue Descartes, the École Polytechnique was run like a military academy. Mandelbrot was issued a uniform, including a "vaguely Napoleonic" two-cornered hat and a century-old straight sword, which he bore in ceremonial marches to honor various dignitaries, among them a visiting Ho Chi Minh. He also received a civil servant's starting pay as pocket money. "This helps answer the question I am often asked by U.S. parents or teachers: 'How come twenty-year-old students in France are so much better in math?' Part of the answer: 'Because they are, in effect, bribed.'"

Upon finishing at Polytechnique, with the equivalent of a master's degree, Mandelbrot made another perplexing (to Uncle Szolem, at least) move. He headed to California, where he intended to study fluid dynamics at Caltech, in Pasadena. Fluid dynamics concerns the motion of a fluid like water or air as it flows around obstacles under the influence of various forces. It is a notoriously hard area of mathematics. And the most intractable problem posed by fluid dynamics is turbulence—the tendency of wind to break into irregular gusts, or of rivers to develop churning eddies. Is there some geometric principle that might explain how a smooth flow abruptly gives way to seemingly unpredictable turbulence? Such a question clearly had the desired Keplerian flavor. Unfortunately, the great authority on fluid dynamics whom he traveled to Pasadena to study with, Theodore von Kármán, turned out to be on leave—in Paris.

On his return to France, Mandelbrot—who by now, because of his Polish nationality and his quasi-military service at Polytechnique, was as unclassifiable civilly as he was mathematically—"fell into the open arms of the French air force." His stint as a draftee at the air force camp at Nanterre (which he calls "Camp de la Folie") was rich in bureaucratic comedy. "I introduced myself to the captain. He was barely five feet tall and hated all six-footers, especially low-ranked ones. He asked to see my papers. 'This letter simply *announces* that you *shall* be appointed. Only the president of France can sign those papers.'" Once the confusion was cleared up, Mandelbrot was chosen by the camp's commander to serve as a scientific liaison to academia. "I liaised with abandon," he says, "and everyone was delighted." At the same time, he began to indulge a passion for classical music, attending recitals in Paris by George Enescu and a "young and skinny (!) flutist named Jean-Pierre Rampal." This passion, once Mandelbrot became famous, led to friendships with conductors like Solti and Abbado and with the composers Ligeti and Charles Wuorinen, whose music came to be influenced by Mandelbrot's notion of fractal self-similarity.

Released from the air force and still short of a doctorate, Mandelbrot, now twenty-six, became a "not-so-young" grad student at the University of Paris, "then at a low point in its long and often glorious history." It was in casting about for a thesis topic that he had his first Keplerian glimmer. One day Uncle Szolem—who by now had written off Mandelbrot as a loss to mathematics—disdainfully pulled from a wastebasket and handed to him a reprint about something called Zipf's law. The brainchild of an eccentric Harvard linguist named George Kingsley Zipf, this law concerns the frequency with which different words occur in written texts—newspaper articles, books, and so on. The most frequently occurring word in written English is "the," followed by "of" and then "and." Zipf ranked all the words in a large variety of written texts in this way and then plotted their frequency of usage. The resulting curve had an odd shape. Instead of falling gradually from the most common word to the least common, as one might expect, it plunged sharply at first and then leveled off into a long and gently sloping tail—rather like the path of a ski jumper. This shape indicates extreme inequality: a few hundred top-ranked words do almost all the work, while the large majority languish in desuetude. (If anything, Zipf underestimated this linguistic inequality: he was using James Joyce's *Ulysses*, rich in esoteric words, as one of his main data sources.) The "law" Zipf came up with was a simple yet precise numerical relation between a word's rank and its frequency of usage.

Zipf's law, which has been shown to hold for all languages, may seem a trifle. But the same basic principle turns out to be valid for a great variety of phenomena, including the size of islands, the populations of cities, the amount of time a book spends on the bestseller list, the number of links to a given website, and—as the Italian economist Vilfredo Pareto had discovered in the 1890s—a country's distribution of income and wealth. All of these are examples of "power law" distributions. (The word "power" here refers not to the political or electrical kind but to the mathematical exponent that determines the precise shape of a given distribution.) Power laws apply, in nature or society, where there is extreme inequality or unevenness: where a high peak (corresponding to a handful of huge cities, or frequently used words, or very rich people) is followed by a low "long tail" (corresponding to a multitude of small towns, or rare words, or wage slaves). In such cases, the notion of "average" is meaningless.

Mandelbrot absorbed Zipf's law on the Métro ride home from his uncle's. "In one of the very few clear-cut eureka moments of my life," he tells us, "I saw that it might be deeply linked to information theory and hence to statistical thermodynamics—and became hooked on power law distributions for life." He proceeded to write his Ph.D. thesis on Zipf's law. Neither his uncle Szolem nor his dissertation committee (headed by Louis de Broglie, one of the founders of quantum theory) paid much heed to his effort to explain the significance of power laws, and for a long time thereafter Mandelbrot was the only mathematician to take such laws and their long tails seriously—which is why, when their importance was finally appreciated half a century later, he became known as the father of long tails.

Having launched himself with his offbeat thesis as a "solo scientist," Mandelbrot sought out other similarly innovative mathematicians. One such was Norbert Wiener, the founder (and coiner) of "cybernetics," the science of how systems ranging from telephone switchboards to the human brain are controlled by feedback loops. Another was John von Neumann, the creator of game theory (and much else). To Mandelbrot, these two men were "made of stardust." He served as postdoc to both: first to Wiener at MIT, and then to von Neumann at the Institute for Advanced Study in Princeton, where he had a nightmarish experience. Delivering a lecture on the deep links between physics and linguistics, he watched as one after another of the famous figures in the audience nodded off and snored. When he finished, the renowned historian of mathematics Otto Neugebauer woke the sleepers by shouting, "I must protest! This is the worst lecture I ever heard." Mandelbrot was by now paralyzed with fear, but happily a formidable pair came to his rescue: first Robert Oppenheimer, who flawlessly conveyed the intended gist of the lecture in one of his legendary "Oppie talks"; and then von Neumann, who did the same in one of his equally celebrated "Johnny talks." The audience was transfixed, and the event ended in triumph.

Returning to Europe, Mandelbrot, now newly married, spent a couple of blissful years with his bride in Geneva. There the psychologist Jean Piaget, impressed by his work in linguistics, tried to take him on as a mathematical collaborator. Mandelbrot declined the offer, despite his (qualified) respect for the great man: "While Piaget could be vague or wrong, he was not a phony." Fernand Braudel invited him to set up a research center in Paris near the Luxembourg Gardens to promote the sort of quantitative history favored by the Annales school.

But Mandelbrot continued to feel oppressed by France's purist mathematical establishment. "I saw no compatibility between a university position in France and my still-burning wild ambition," he writes. So, spurred by the return to power in 1958 of Charles de Gaulle (for whom Mandelbrot seems to have had a special loathing), he accepted the offer of a summer job at IBM in Yorktown Heights, north of New York City. There he found his scientific home.

As a large and somewhat bureaucratic corporation, IBM would hardly seem a suitable playground for a self-styled maverick. The late 1950s, though, were the beginning of a golden age of pure research at IBM. "We can easily afford a few great scientists doing their own thing," the director of research told Mandelbrot on his arrival. Best of all, he could use IBM's computers to make geometric pictures. Programming back then was a laborious business that involved transporting punch cards from one facility to another in the backs of station wagons. When his son's high school teacher sought help for a computer class, Mandelbrot obliged, only to find that soon students all over Westchester County were tapping into IBM's computers by using his name. "At that point, the computing center staff had to assign passwords," he says. "So I can boast—if that's the right term—of having been at the origin of the police intrusion that this change represented."

It was chance again that led to Mandelbrot's next breakthrough. Visiting Harvard to give a lecture on power laws and the distribution of wealth, he was struck by a diagram that he happened to see on a chalkboard in the office of an economics professor there. The diagram was almost identical in shape to the one Mandelbrot was about to present in his lecture, yet it concerned not wealth distribution but price jumps on the New York Cotton Exchange. Why should the pattern of ups and downs in the market for cotton bear such a striking resemblance to the wildly unequal way wealth was spread through society? This was certainly not consistent with the orthodox model of financial markets, which was originally proposed in 1900 by a French mathematician named Louis Bachelier (who had copied it from the physics of a gas in equilibrium). According to the Bachelier model, price variation in a stock or commodity market is supposed to be smooth and mild; fluctuations in price, arranged by size, should line up nicely in a classic bell curve. This is the basis of what became known as the efficient market hypothesis.

But Mandelbrot, returning to IBM and sifting with the aid of its computers through a century of data from the New York Cotton Exchange, found a far more volatile pattern, one dominated by a small number of extreme swings. A power law seemed to be at work. Moreover, financial markets behaved roughly the same on all timescales. When Mandelbrot took a price chart and zoomed in from a year to a month to a single day, the wiggliness of the line did not change. In other words, price histories were self-similar—like a cauliflower. "The very heart of finance," Mandelbrot concluded, "is fractal."

The fractal model of financial markets that Mandelbrot went on to develop has never caught on with finance professors, who still by and large cling to the efficient market hypothesis. If Mandelbrot's analysis is right, reliance on orthodox models is dangerous. And so it has proved, on more than one occasion. In the summer of 1998, for example, Long-Term Capital Management—a hedge fund founded by two economists who had been awarded Nobel Prizes for their work in portfolio theory and staffed with twenty-five Ph.D.'s—blew up and nearly took down the world's banking system when an unforeseen Russian financial crisis foiled its models.

Mandelbrot resented being "pushed out of the economic mainstream." He recounts, with some bitterness, how a job offer from the University of Chicago's business school, a bastion of efficient market orthodoxy, was extended to him and then rescinded by its dean (and Reagan's future secretary of state), George Shultz. Harvard, too, declined to offer a permanent position to the visiting Mandelbrot after initially expressing interest. He took these seeming snubs in stride. Upon returning to IBM, he experienced "the warm feeling of coming home to the delights of old-fashioned collegiality in a community far more open and 'academic' than Harvard." Indeed, Mandelbrot was to remain based at IBM until 1987, when the corporation saw fit to end its support for pure research. He was thereupon invited to teach at Yale, where in 1999, at the age of seventy-five, he finally got academic tenure—"in the nick of time."

It was at Harvard in 1980 that Mandelbrot made the most momentous discovery of his career. Through his friend Stephen Jay Gould—a fellow champion of the idea of discontinuity—Mandelbrot was invited to teach a course showing how fractal ideas could shed light on classical mathematics. This led him to take up "complex dynamics," an abstract approach to the study of chaos. Complex dynamics had flourished in Parisian mathematical circles in the early twentieth century, but it soon led to geometric forms that were far too complicated to be visualized, and the subject became frozen in time.

Mandelbrot saw a way to unfreeze it—through the power of the computer. At the time, computers were pretty well disdained by mathematicians, who "shuddered at the very thought that a machine might defile the pristine 'purity' of their field." But Mandelbrot, never a purist, got his hands on a brand-new VAX supermini in the basement of Harvard's science center. Using its graphic capabilities as a sort of microscope, he set out to investigate a certain geometric figure generated by a very simple formula (the only formula, by the way, he allowed to appear in his memoir). What he found, as the computer produced increasingly detailed pictures of this figure, was utterly unexpected: a wondrous world of beetle-shaped blobs surrounded by exploding buds, tendrils, curlicues, stylized sea horses, and dragon-like creatures, all bound together by a skein of rarefied filaments. At first he suspected that the geometric riot he was seeing was due to faulty equipment. But the more the computer zoomed in, the more precise (and fantastic) the pattern became; indeed, it could be seen to contain an infinite number of copies of itself, on smaller and smaller scales, each fringed with its own rococo decorations. This was what came to be dubbed the Mandelbrot set.

Mandelbrot justly deems the set named in his honor to be a thing of "infinite beauty." Its detailed geometry, still far from fully understood, encodes an infinite bestiary of chaotic processes. How could such an endlessly complex object—the most complex, it has been claimed, in all mathematics—arise from such a simple formula? For the mathematical physicist Sir Roger Penrose, this uncovenanted richness is a striking example of the timeless Platonic reality of mathematics. "The Mandelbrot set is not an invention of the human mind: it was a discovery," Penrose has written. "Like Mount Everest, the Mandelbrot set is just there!"

The Fractalist is not a flowing memoir; indeed, it has a fractal roughness of its own. No doubt it would have been more polished had the author lived longer: his passion for revising and tinkering with sloppy drafts, he tells us, rivaled Balzac's. Occasionally, there are jarring notes of hauteur ("I reach beyond arrogance when I proclaim . . .") and injured merit ("I don't seek power or run around asking for favors . . . Academia found me unsuitable"). Little attempt is made to explain to the lay reader the author's mathematical innovations, like his use of dimension to measure fractal roughness. (The coastline of Britain, for example, is so wiggly that it has a fractal dimension of 1.25; it thus falls somewhere between a smooth curve, which has dimension 1, and a smooth surface, which has dimension 2.)

Mandelbrot can be pardoned for not belaboring such technicalities. His tone as a memoirist is more philosophical. The world we live in, he observes, is an "infinite sea of complexity." Yet it contains two "islands of simplicity." One of these, the Euclidean simplicity of smooth forms, was discovered by the ancients. The other, the fractal simplicity of self-similar roughness, was largely discovered by Mandelbrot himself. His geometric intuition enabled him to detect a new Platonic essence, one shared by an unlikely assortment of particulars, ranging from the humble cauliflower to the sublime Mandelbrot set. The delight he took in roughness, brokenness, and complexity, in forms that earlier mathematicians had regarded as "monstrous" or "pathological," has a distinctly modern flavor. Indeed, with their intricate patterns that recur endlessly on ever tinier scales, Mandelbrot's fractals call to mind the definition of beauty offered by Baudelaire: "C'est l'infini dans le fini."

PART IV

Higher Dimensions, Abstract Maps

9

Geometrical Creatures

One feature of the world that few people stop to puzzle over is how many dimensions it has. Although it is a little tricky to say just what a dimension is, it does seem fairly obvious that we, the objects that surround us, and the space we move about in are structured by three dimensions, conventionally referred to as height, breadth, and depth. Even philosophers have tended to take this for granted. Aristotle, at the outset of *On the Heavens*, declared that "the three dimensions are all there are." Why? Because, he argued in a somewhat mystical vein, the number 3 comprises beginning, middle, and end; therefore, it is perfect and complete. A less mystical proof of nature's three-dimensionality was supplied by the Alexandrian astronomer Ptolemy. Once you have arranged three sticks so that they are mutually perpendicular and meet at a point, Ptolemy observed, it is impossible to add a fourth; therefore, further dimensions are "entirely without measure and without definition." Ptolemy's reasoning was later ratified by Galileo and by Leibniz, who declared the three dimensions of space to be a matter of geometric necessity.

The first philosophers to talk about a "fourth dimension" were the seventeenth-century Cambridge Platonists, but they seemed to have something more spiritual than spatial in mind. One of them, Henry More, suggested in 1671 that the fourth dimension was the abode of Platonic ideas and, quite possibly, ghosts. Around the same time, Descartes took the seemingly innocuous step of adding an extra variable to his coordinate geometry, which enabled him to define four-dimensional *sursolides*. Timid contemporaries found this intolerable; in 1685, the mathematician John Wallis denounced the *sursolide* as a "Monster in Nature, less possible than a Chimera or Centaur!"

Kant, in his early writings at least, flirted with the idea that three-dimensional space might be contingent; perhaps, he conjectured, God created other worlds with different numbers of dimensions. By the time he wrote the *Critique of Pure Reason*, however, Kant had decided that space was not an objective feature of reality but something imposed on it by the mind to give order to experience. Moreover, he held, its character was irrevocably Euclidean and three-dimensional; this we know with "apodictic certainty." In 1817, Hegel asserted, without much in the way of proof, that the necessity of three dimensions was founded in the very nature of the notion of space.

Meanwhile, in the world of mathematics, a revolution was getting under way. During the first decades of the nineteenth century, Gauss, Lobachevsky, and Bolyai were independently exploring "curved" geometries, where the shortest distance between two points was no longer a straight line. In the 1840s, Arthur Cayley and Hermann Grassmann, also working independently, extended the Euclidean framework to spaces of more than three dimensions. These developments were brought together in a magnificent synthesis by Bernhard Riemann (1826–1866). In a lecture before the faculty of the University of Göttingen on June 10, 1854, titled "On the Hypotheses Which Lie at the Foundation of Geometry," Riemann toppled the Euclidean orthodoxy that had dominated mathematics—and, indeed, Western thought—for two millennia. According to Euclid, a point has zero dimensions; a line, one; a plane, two; and a solid, three. Nothing could have four dimensions. Moreover, Euclidean space is "flat": parallel lines never meet. Riemann transcended both these assumptions, rewriting the equations of geometry so that they could describe spaces with any number of dimensions and with any kind of curvature. In doing so, he defined a collection of numbers, called a tensor, that characterizes the curvature of a higher-dimensional space at each point.

Riemann's *n*-dimensional non-Euclidean geometry was a pure intellectual invention, unmotivated by the needs of contemporary science. Six decades later, his tensor calculus furnished precisely the apparatus that Einstein required to work out the general theory of relativity. But the immediate effect of Riemann's revolution was to destroy the old notion of geometry as the science of physical space. Clearly, there was nothing metaphysically necessary about three dimensions. An endless variety of other spatial worlds was possible; they could be described by a logically consistent theory, and hence were mathematically real. This led to an interesting line of speculation. Could such worlds ever be visualized by the human imagination? What would it be like to exist in one? Was it just an accident that among all the possible spatial architectures we find ourselves living in a world of three dimensions? Or—an even more heady thought—could it be that we *do* live in a world with more dimensions, but like the prisoners in Plato's cave we are too benighted to realize it?

"A person who devoted his life to it might *perhaps* eventually be able to picture the fourth dimension," wrote Henri Poincaré in the late nineteenth century. He made the task seem rather daunting, but maybe that is because, as a mathematician of peerless spatial intuition himself, he had very high standards. Others felt that some intuition of the fourth dimension might be attainable with a more modest expenditure of time. In an 1869 address to the British Association, titled "A Plea for the Mathematician," James J. Sylvester argued that it was time for higher-dimensional geometry to come out of the closet; with a little practice, Sylvester maintained, it was perfectly possible to visualize four dimensions. Some nonmathematical charlatans went even further. In 1877, the fourth dimension achieved notoriety when an American psychic named Henry Slade was put on trial in London for fraud. In a series of séances with prominent members of London society, Slade had purported to summon spirits from the fourth dimension. Distinguished physicists, including two future Nobel laureates, sprang to Slade's defense, gulled by his knack for doing parlor tricks that supposedly exploited this unseen extension of space (like miraculously extracting objects from sealed three-dimensional containers). Although Slade was convicted, the mysterious fourth dimension had captured the public imagination. It was in this atmosphere that a Victorian schoolmaster named Edwin A. Abbott wrote the first, and most enduringly successful, popular fiction on the subject, a little masterpiece titled *Flatland: A Romance of Many Dimensions*.

Since its initial publication in 1884, *Flatland* has been through countless editions, with introductions by Ray Bradbury and Isaac Asimov,
among others. The most rewarding of them, to my mind, is the 2002
one annotated by Ian Stewart: *The Annotated Flatland*. Why an annotation? Because, as its subtitle suggests, *Flatland* itself has several
dimensions: it is a scientific fantasy that coaxes the reader into imagining unseen spatial realms; it is a satire on Victorian attitudes, especially those concerning women and social status; it is an allegory of a
spiritual journey.

Edwin Abbott was a typically tireless Victorian, rating two double-
columned pages in the *Dictionary of National Biography*. He was the
longtime headmaster of the City of London School, where his devoted
pupils included the future prime minister H. H. Asquith. He was a
Broad Church reformer, celebrated for his preaching at Oxford and
Cambridge. An avid Shakespeare scholar, he produced *A Shakespear-
ian Grammar* (1870), which became a reference standard, along with
some fifty other books, many on ponderous theological matters. What
possessed him to write a consciousness-raising mathematical satire like
Flatland—his only book still in print—is not really known. As a progressive educator, he might have wished to shake up the English mathematics curriculum, with its dreary emphasis on the memorization of
long proofs from Euclid. And, as a modern-minded churchman, he
was no doubt attracted to the challenge of reconciling the spiritual and
scientific worldviews.

Abbott's doorway to higher dimensions was the method of analogy. We cannot readily picture to ourselves a space with one more dimension than our three-dimensional world. We can, however, imagine
a space with one *less* dimension: a plane. Suppose there were a society
of two-dimensional creatures confined to a planar world. What would
they look like to us? And—a more interesting question—how would
three-dimensional beings like us look to them, assuming we could somehow pass through their world or lift them into ours?

Abbott's Flatland consists of an infinite plane inhabited by a rigidly
hierarchical society of geometric creatures. Social pedigree is determined by the number of sides the creature has: lowly women are mere
line segments, whereas men range from working-class triangles to
bourgeois squares to aristocratic polygons with five sides and up; the

priestly caste consists of polygons with so many sides they are virtually circular. There is some upward mobility: in a caricature of Victorian notions of progressive evolution, well-behaved members of Flatland's lower classes occasionally bear children with a greater number of sides than themselves. Irregularity of form is equated with moral obliquity and criminality, and eugenic infanticide is practiced on newborns "whose angle deviates by half a degree from the correct angularity."

The narrator of the book is a conservative lawyer aptly named A. Square. In decorous language, he explains to the reader (presumed to be a three-dimensional "Spacelander" like us) the architecture of his world, its history, politics, and folkways. Flatland abounds in absurdities, which are often imperfectly appreciated by Mr. Square. For example, women, though "devoid of brainpower," are extremely dangerous because of their needlelike linear form; in a fit of anger (or sneezing) the "Frail Sex" is capable of inflicting instant death by penetrating a polygonal male.

Life in a two-dimensional world also poses many practical difficulties. How can Flatlanders possibly recognize one another by sight? Suppose you are looking at a penny on a table. Viewed from above, its round shape is apparent. But if you get down to the level of the tabletop, the penny's edge appears to be a straight line. In Flatland, everyone and everything looks like a straight line; Flatlanders cannot "see" angles. To overcome the recognition problem, women and tradesmen feel one another in their social intercourse. ("Permit me to ask you to feel and be felt by my friend Mr. So-and-so.") The upper classes find this beyond vulgar; "to *feel* a Circle would be considered a most audacious insult," Mr. Square tells us. Instead, they rely on their cultivation of depth perception and geometry to determine one another's angles, and hence social rank.

At one point in Flatland's history, reformers attempted to eliminate the invidious distinction between the feeling and the non-feeling classes by introducing a "Universal Colour Bill." Henceforth, Flatlanders would have the right to adorn their linear profiles with colors of their choice, thus achieving equality of recognition and, incidentally, relieving the aesthetic dullness of their world. But the popular uprising in favor of this reform was bloodily suppressed, and the forces of reaction banished color from Flatland.

The social satire in Flatland tends to be schematic and a bit labored. Nor is Abbott altogether successful in setting up a consistent scheme for two-dimensional life. Take the question of how sound and hearing work in Flatland. In a space with an odd number of dimensions, like our own three-dimensional world, sound waves move in a single sharp wave front. If a gun is fired some distance away, you hear first silence, then a bang, then silence again. But in spaces with an even number of dimensions, like a two-dimensional plane, a noise-like disturbance will generate a system of waves that reverberate forever.

Another matter overlooked by Abbott is how a Flatlander's brain might work. Imagine the nightmare of having to draw up neural circuitry on a two-dimensional piece of paper, where there is no room for wires to cross each other without intersecting. In his annotation of *Flatland*, Ian Stewart cleverly solves this problem for the author by invoking the theory of "cellular automata": two-dimensional arrays of cells that, communicating with their neighbors according to simple rules, can carry out any task that a computer could accomplish. With cellular automata for brains, Flatlanders would be capable of intelligent behavior (although, as philosophers like John Searle would argue, they might fall short of consciousness). As for more intimate details of the Flatlanders' existence—their method of sexual congress, for example—the author maintains a proper Victorian reticence.

But Abbott is not primarily interested in the details of a functioning two-dimensional world. His real theme is *higher* dimensions, and that is taken up in the second (and more interesting) half of the book. If part 1 of *Flatland* is a swipe at Victorian society, part 2 is a swipe at those who refuse to admit the possibility of unseen worlds.

One night, while Mr. Square is at home with his wife, he is visited by a ghostly "Stranger," a spherical being from Spaceland. How can the three-dimensional sphere enter a two-dimensional world like Flatland? The reader is invited to picture Flatland as something like the surface of a pond. By rising from the deep and breaking this surface, the sphere can manifest himself to the two-dimensional beings who float upon it. At first, they would see nothing at all. Then, as the sphere first made contact with the surface, they would see a single point. As the sphere continued to rise, they would see this point expand into a circle, whose radius would grow until it reached a maximum when

the sphere had passed halfway through the surface. Then the Flatlanders would see the circle begin to contract, shrinking to a point again
and then disappearing altogether as the sphere rose completely above
the surface.

Mr. Square is alarmed by this apparition. How can the Stranger—
who must be of the priestly caste, judging from his seeming circularity—
appear from nowhere, make himself smaller or larger at will, and then
magically vanish? The Stranger explains that "in reality I am not a Circle, but an infinite number of Circles," all of varying sizes. Mr. Square,
whose imagination cannot grasp the idea of a three-dimensional solid, is
incredulous. So the Stranger tries to prove his assertion with several
tricks, like rising up out of Flatland, hovering invisibly over Mr. Square,
and touching the center of his stomach. Finally, the Stranger lifts
the enraged Mr. Square out of Flatland into the world of space. There,
floating as insubstantially as a sheet of paper in the breeze, Mr. Square
looks down on his two-dimensional world and sees the entire form
and interior of every person and building. (To return to the example of
the penny on the table, imagine looking at its edge from the level of
the tabletop and then rising up and looking at it from above: suddenly
you can see the whole circular form and Lincoln's head "inside" it.)
Even more awesome is Mr. Square's vision of the Stranger, who now
appears in his full tridimensionality: "What seemed the centre of the
Stranger's form lay open to my view: yet I could see no heart, nor
lungs, nor arteries, only a beautiful harmonious Something—for which
I had no words; but you, my Readers in Spaceland, would call it the
surface of the Sphere."

Now entertain, if you will, the higher-dimensional analogue of
all this. What would it look like if a four-dimensional Stranger—a
"hypersphere"—were to pass through our three-dimensional Spaceland? First we would see nothing; then a point-like ball would appear,
expand into a sphere, then dwindle to a point and vanish. The spheres
of varying sizes that we saw would be 3-D cross sections of the hypersphere, just as the circles that Mr. Square saw were 2-D cross sections
of the sphere. Simple enough. The hard part is getting your imagination to regard these three-dimensional appearances as the gradual
revelation of a four-dimensional entity. What would we see if, like
Mr. Square, we were actually yanked out of our world into "a more

spacious Space" with an extra dimension? No doubt we would be as dumbstruck by the vision of the hypersphere as Mr. Square was by that of the sphere. We would also be astonished to find, looking back upon our 3-D world, that all objects were transparent to our sight and visible from every perspective simultaneously. We could reach inside someone's body and remove his appendix without breaking his skin (a great boon for surgeons). We could pick up a left shoe, rotate it in the fourth dimension, and restore it to the 3-D world as a right shoe.

Once Mr. Square gets habituated to Spaceland, he is quick to grasp the analogy to higher dimensions. If his familiar 2-D world is an infinitesimal sliver of 3-D space, he reasons, perhaps 3-D space is but a sliver of a 4-D space, a space harboring hyperspheres that surpass the spherical Stranger the way the Stranger surpasses the circular priests of Flatland. And why stop there? "In that blessed region of Four Dimensions, shall we linger on the threshold of the Fifth, and not enter therein?" he rhetorically asks of the Stranger. "Ah, no! Let us rather resolve that our ambition shall soar with our corporal ascent. Then, yielding to our intellectual onset, the gates of the Sixth Dimension shall fly open; after that a Seventh, and then an Eighth . . ."

But the Stranger reacts coldly to Mr. Square's ecstatic aspirations. Ironically, although the Stranger has come to preach the gospel of three dimensions, he himself cannot make the imaginative leap into the fourth dimension. "Analogy!" he harrumphs. "Nonsense: what analogy?" In a fit of dudgeon, he hurls Mr. Square back into Flatland, where the authorities promptly clap our narrator into prison for his subversive ravings about a higher reality.

The spiritual aspect of Flatland is patent. It tantalizes the reader with the possibility that there is an unseen world surrounding us, one that lies in an utterly novel direction—"Upward, but not Northward!" as Mr. Square tries to express it—a world that contains miraculous beings which can hover about us without actually occupying our own space. Soon English clergymen were talking about the fourth dimension as the abode of God and his angels. The same year Flatland was published, a pamphlet called *What Is the Fourth Dimension?* appeared in England, bearing the subtitle "Ghosts Explained." Its author, Charles Hinton, was a tireless evangelist of the fourth dimension and a self-styled lothario; "Christ was the Saviour of men, but I am the savior of women, and I don't envy Him a bit!" was one of his sayings. After wedding Mary Boole (the eldest daughter of George Boole, inventor of Boolean algebra), Hinton also decided to marry one of his mistresses. Convicted of bigamy, he left England and ended up in the United States, where he became a mathematics instructor at Princeton. ("Here," writes Stewart in his annotation, "he invented a baseball pitching machine that propelled the balls via a charge of gunpowder. It was used for team practice for a while, but it proved a little too ferocious, and after several accidents it was abandoned.")

In 1907, Hinton wrote a sort of sequel to Abbott's book, which he titled *An Episode of Flatland*. Although it failed to achieve the success of *Flatland*, Hinton did surpass Abbott in finding ways to visualize four-dimensional objects. Abbott, as we saw, exploited the method of cross sections: one imagines a 4-D object by looking at its 3-D cross sections as it traverses 3-D space. Hinton added to this the "method of shadows," whereby one tries to grasp a 4-D object by considering the 3-D shadows, or projections, it casts from various angles. One projection of a hypercube, for example, looks like a small 3-D cube inside of a larger one—an object that Hinton dubbed a "tesseract." Finally, there is the "method of unfolding." Just as you can unfold a 3-D cardboard box into a flat cross made up of six squares, you can also "unfold" a 4-D hypercube into a 3-D cross shape consisting of eight cubes. Salvador Dalí depicts such an unfolded 4-D cube in his crucifixion painting *Corpus Hypercubus*, which is in the collection of New York's Metropolitan Museum of Art.

Thanks to the efforts of Abbott and Hinton, and to the popular writings of Henri Poincaré in France, the "fourth dimension" had become a household expression by the early twentieth century. It cropped up in the works of authors like Alfred Jarry, Proust, Oscar Wilde, and Gertrude Stein. (In 1901, Joseph Conrad and Ford Madox Ford cowrote a novel called *The Inheritors* about a race of beings from the fourth dimension who take over our world.) It seduced the artistic avant-garde, who invoked it to justify the overthrow of three-dimensional Renaissance perspective. The cubists were especially enthralled by the idea: from the fourth dimension, after all, a three-dimensional object or person could be seen from all perspectives at once. (Recall how Mr. Square, on ascending above Flatland, could for the first time see all the edges of the objects in that two-dimensional world.) Apollinaire wrote in *Les peintres cubistes* (1913) that the fourth dimension "is space itself" and "endows objects with plasticity."

The notion of unseen higher dimensions was also seized upon by Theosophists, who saw it as a weapon against the evils of scientific positivism and who cultivated their "astral vision," the better to apprehend it. The mystic P. D. Ouspensky introduced the fourth dimension to tsarist Russia as the solution to all the "enigmas of the world." Vladimir Lenin, alarmed by the seeming spiritualist implications of the fourth dimension, attacked the notion in his *Materialism and Empiriocriticism* (1909). Mathematicians might explore the possibility of four-dimensional space, Lenin wrote, but the tsar can only be overthrown in the three-dimensional world.

Curiously, the one part of culture that was largely untouched by the craze over higher dimensions was science. Not only was the idea rendered disreputable by its association with mystics and charlatans; it also seemed devoid of testable consequences. Then, during World War I, Einstein framed his general theory of relativity. By uniting the three dimensions of space and the one dimension of time into a four-dimensional manifold, "space-time," and then explaining gravity as a curvature within this manifold, Einstein's theory gave the impression that the hitherto mysterious fourth dimension was merely time. (This impression was a wrongheaded one, because according to relativity there is no privileged way of slicing up the four-dimensional space-time manifold into purely spatial and temporal dimensions.) When newspaper headlines announced the triumphant confirmation of general relativity in 1919, the idea that time was the fourth dimension entered the culture at large, and interest in higher spatial dimensions began to dry up—quite prematurely, as we shall see.

To mathematicians, of course, it is irrelevant how many dimensions the physical space we live in happens to have. The non-Euclidean revolution of the mid-nineteenth century freed them to explore the spatial structure not just of the actual world but of all conceivable worlds. And this they continued to do even after the passing of the fourth-dimension vogue, inventing ever more exotic varieties: "Hilbert space," with an infinite number of dimensions; "fractal" spaces, which might have, say, two and a half dimensions; rubbery topological spaces; and on and on. Geometry has come a long way since *Flatland*.

The space we actually live in might seem boring in comparison with the rococo spaces of higher mathematics. Around four decades ago, however, physicists were forced to consider the possibility that there might be more to our spatial world than meets the eye, dimensionally speaking. To understand why, consider that contemporary physics has two sets of laws: general relativity, which describes how things behave on a very massive scale (stars on up); and quantum theory, which describes how things behave on a very small scale (atoms on down). That may seem like a nice division of labor. But what happens when you want to describe something that is both very massive and very small—like the universe a split second after the big bang? Somehow general relativity and quantum theory must be fitted together into a theory of everything. However, it seems impossible to do this for a world that has only three spatial dimensions. The only known way to make relativity theory consistent with quantum theory is by supposing that the basic objects that make up our universe are not one-dimensional particles but two-dimensional strings and still-higher-dimensional "branes" (a term derived from "membrane"). Moreover, if the unified theory—called string theory, or sometimes M-theory—is to be mathematically coherent, these strings and branes must be vibrating in a space that has no fewer than nine dimensions.

String theory, then, requires that the universe have six dimensions beyond the three we are familiar with. Why don't we see them? There are two hypotheses. One, long favored by string theorists, is that the six surplus dimensions are "compactified"—that is, curled up into circles of vanishingly small radius. (Think of a garden hose: from a distance it looks like a one-dimensional line, but on closer inspection it has a tiny circular dimension too.) More recently, though, physicists have speculated that the additional dimensions might be of macroscopic scale, or even infinite in extent. Yet we fail to notice them because all the particles that make up our familiar world are stuck to a three-dimensional membrane that is adrift in the higher-dimensional world. If this proves true, we are in a situation very similar to that of the Flatlanders. But whereas Mr. Square needed a visit from a Stranger to realize that higher dimensions existed, we have arrived at this hypothesis by theoretical reasoning and may even be able to confirm it experimentally. (This could be done, for instance, by smashing together subatomic particles and seeing whether any of the new particles produced in the collision disappeared into an extra dimension.)

If string theory's implication that we are surrounded by unseen dimensions pans out—a big "if," by the way—this would constitute yet another Copernican revolution in how we understand our place in the scheme of things. As the physicist Nima Arkani-Hamed has put it, "The earth is not the center of the solar system, the sun is not the center of our galaxy, our galaxy is just one of billions in a universe that has no center, and now our entire three-dimensional universe would be just a thin membrane in the full space of dimensions. If we consider slices across the extra dimensions, our universe would occupy a single infinitesimal point in each slice, surrounded by a void."

There remains one important question unaddressed by Edwin Abbott in *Flatland*. It is the question implicitly posed (if not satisfactorily answered) by Aristotle: Why does our everyday world have three dimensions? Since the middle of the nineteenth century, it has been known that this is not a matter of geometric necessity. Can science then provide an explanation, or is it just a cosmic accident?

String theorists have come up with some extremely elegant and subtle conjectures as to why, out of the nine spatial dimensions they posit, exactly three of them expanded to enormous size after the big bang while the remaining six got choked off and remained tiny. But there is another sort of explanation, one that is perhaps easier to grasp: in a world where the number of spatial dimensions was anything other than three, beings like us simply could not exist. In a space of more than three dimensions, there would be no stable planetary orbits. (This was proved a century ago by Paul Ehrenfest.) Nor would there be stable orbits for electrons within atoms. Therefore, there could be no chemistry, and hence no chemically based life-forms, in a world of more than three spatial dimensions.

Well, then, what about a world of *fewer* than three spatial dimensions? As noted earlier, sound waves would not propagate cleanly in 2-D Flatland, or indeed in any other space with an even number of dimensions. But the difficulties are not limited to sound; it is impossible to transmit well-defined signals of any type in a space that has an even number of dimensions. This rules out the kind of information processing that is essential to intelligent life. So, by a process of elimination (it is left as an exercise for the reader to work out why intelligent life could not exist in a one-dimensional "Lineland"), the only kind of world that is congenial to chemically based, information-processing beings like us is one with precisely three spatial dimensions. So it should not come as a surprise that we find ourselves living in a three-dimensional world. (Physicists call this "anthropic" reasoning.)

And we should not repine. In a fundamental sense, three-dimensional space is the richest space of all. Clearly, it beats a space of one or two dimensions: in Lineland and Flatland, there is no room for any interesting complexity. (Recall how visually impoverished life is in Flatland, where everything looks like a line segment.) As for spaces of four dimensions and up, they are too "easy": there are so many degrees of freedom, so many options for rotating things and moving them around, that complexities are readily rearranged and dissolved. Only in three-dimensional space do you get the right creative tension—which may be why mathematicians find it the most challenging of all.

Take Poincaré's conjecture, among the greatest (and stubbornest) problems in modern mathematics. Basically, it asserts that any blob in n dimensions that has a certain algebraic property can be massaged into an n-dimensional sphere. Poincaré framed this conjecture in 1904. By 1961, it was shown to hold in any space of five dimensions or more. In 1982, it was proved true in four-dimensional space. But only in the present century was Grigori Perelman able to resolve what proved to be the trickiest case of Poincaré's conjecture: the case of three dimensions.

Training ourselves to conceive of "a more spacious Space" than that of our three-dimensional world may enlarge our imagination and contribute to the progress of science. And we can surely sympathize with the aspiration of Mr. Square—not to mention assorted Theosophists, Platonists, and cubists—to rise up into the splendor of the fourth dimension and beyond. But we need not follow them. For intellectual richness and aesthetic variety, a world of three dimensions is world enough.

10

A Comedy of Colors

A century and a half ago, a student who was coloring a map of England noticed that he only needed four colors to do the job—that is, to ensure that no counties sharing a border, such as Kent and Suffolk, got the same color. This led him to guess that four colors might be sufficient for any map, real or invented. He mentioned this idle surmise to his brother. His brother in turn mentioned it to a distinguished mathematician, who, after a little experimentation to see if it looked plausible, tried and failed to prove that it was true.

In the decades that followed, many other mathematicians, along with innumerable amateurs—including a great French poet, a founder of American pragmatism, and at least one bishop of London—were similarly engrossed and confounded by the map problem. So easy to state that a child can understand it, the "four-color conjecture" came to rival Fermat's last theorem as the most famous conundrum in all mathematics. Finally, in 1976, the world received the news that the conjecture had been resolved. When it transpired how this was accomplished, however, the celebratory mood gave way to disappointment, skepticism, and outright rejection. What had been a problem in pure mathematics evolved into a philosophical question, or rather a pair of them: How do we justify our claims to mathematical knowledge? And can machine intelligence help us grasp a priori truths?

For all of its mathematical and philosophical interest, the map problem has no obvious practical import—at least not for mapmakers, who show no tendency to minimize the number of colors they use. Still, it is instructive to approach the matter by way of a glance at an actual atlas. Turn to a map of Europe, and look at the part consisting of Belgium, France, Germany, and Luxembourg. Each of these countries shares a border with the other three, so it is pretty obvious that they cannot be distinguished with fewer than four colors. You might think that four colors would be needed only when a map contains a quartet of mutually neighboring regions like this. If you do, turn to a map of the United States, and look at Nevada along with the five states that ring it (California, Oregon, Idaho, Utah, and Arizona). No four of these states are mutually neighboring, the way Belgium, France, Germany, and Luxembourg are. Yet the cluster as a whole cannot be distinguished with fewer than four colors, as you can easily verify. On the other hand—and this may shake your intuition a bit—Wyoming and the six states that ring it can be distinguished with a mere *three* colors.

Some maps need four colors: that much is patent. What the four-color conjecture asserts is that there is no possible map that needs more than four colors. What would it mean to "resolve" this conjecture?

There are two possibilities. Suppose—as some mathematicians have believed—the conjecture is false. Then drawing just one map that required five or more colors would clinch the matter. (In the *Scientific American* of April 1975, Martin Gardner published a complicated map, consisting of 110 regions, that he claimed could not be colored with fewer than five colors. Hundreds of readers sent in copies of the map that they had laboriously colored with just four colors, perhaps not realizing that Gardner was enjoying a little April Fools' joke.) To establish that the four-color conjecture is true, by contrast, would mean showing that every conceivable map, of which there are infinitely many, could be colored with only four colors, no matter how numerous and convoluted and gerrymandered its regions might be.

So the simplicity of the four-color conjecture is deceptive. Just how deceptive is made clear by a look at the long history of the quest to resolve it—a comedy of errors. Francis Guthrie, the student in London who in 1852 first guessed that four colors sufficed, apparently imagined he had proved the conjecture as well. Although Guthrie later became a professor of mathematics in South Africa, he never published anything on the map problem, preferring, it seems, to indulge his interest in botany (a species of heather is named after him). He did, however, discuss it with his younger brother Frederick, who brought it to the attention of Augustus De Morgan, his mathematics professor. De Morgan was a pretty good mathematician and an important figure in the development of logic. Drawn in by the four-color problem, he grew obsessed with the idea that if a map contained four mutually neighboring regions, one of them must be completely enclosed by the other three (as, to revert to an earlier example, Luxembourg is completely enclosed by Belgium, France, and Germany). He believed, quite wrongly, that this "latent axiom" was the key to proving the conjecture, which he continued to vex over until his death in 1871.

It was De Morgan who first mentioned the four-color conjecture in print—in, of all places, an unsigned philosophy review he contributed in 1860 to a popular literary journal, *The Athenaeum*. Word of it crossed the Atlantic to the United States, where the philosopher C. S. Peirce fell under its spell. Peirce pronounced it "a reproach to logic and to mathematics that no proof had been found of a proposition so simple," and he presented his own would-be proof to a mathematical society at Harvard in the late 1860s. No record of it survives. Later, however, Peirce had to acknowledge another man's breakthrough, which he helped publicize in a notice that appeared in *The Nation* on Christmas Day 1879. In doing so, Peirce unwittingly ratified what would become the most famous fallacious proof in the history of mathematics.

Here is perhaps a good place to say something about how mathematicians go about proving propositions—especially those, like the four-color conjecture, that encompass an infinite number of cases. One approach is the method of mathematical induction. This is sometimes likened to knocking down an endless row of dominoes. The crucial step in mathematical induction is showing that if such and such is true for the number n, then it is also true for $n + 1$. In the domino simile, that means each falling domino knocks down its immediate successor, ensuring that all of them ultimately fall. To apply mathematical induction to the map problem, one would have to show that if any map having n regions can be colored with four colors, then so can any map having $n + 1$ regions. That turns out to be terribly tricky to do. When you add the $(n + 1)$th region to a given map, you might have to recolor some or all of the n other regions to make the new one fit in the four-color scheme. No one has ever been able to find a general recipe for such a re-coloration. So much for the falling-dominoes approach.

Happily, there is another strategy for proving a proposition of infinite scope: proof by contradiction. You take the denial of what you want to establish and show that it leads to absurdity. In the case of the four-color conjecture, this means you pretend that there exist counterexamples to the conjecture—maps that need five or more colors—and then derive a contradiction from that assumption. Because such counterexamples would violate the four-color principle, they are colloquially called criminals. Criminal maps, if they existed, could contain any number of regions, but it is useful to focus on those having the absolute smallest number. These are called "minimal" criminals. (Obviously, a minimal criminal would have to have at least five regions to need five colors.) Any map with fewer regions than a minimal criminal is bound, by definition, to be law-abiding—that is, four-colorable.

Now we are in an interesting position. Take a map that is supposedly a minimal criminal. Pick one of its regions, and shrink that region down to a point. This reduces the number of regions by one. So the reduced map doesn't have enough regions to be a criminal (because it has one less region than the minimal criminal we started with). Therefore, the reduced map must be law-abiding—which means it can be colored with four colors. So color it.

Now reverse the process. Restore the region you previously shrank to a point by blowing it up again. This gives you back the original map, with every region appropriately colored except the one that was shrunken and restored. Now ask this: Is there any way to color the restored region so that it fits into the four-color scheme that you applied to the reduced map? Well, that depends on which region you originally picked for the shrink-and-restore move. If, for instance, the region you picked bordered on only three other regions, then you are in luck: When you restore it, there will be a color left over in the four-color scheme that distinguishes the restored region from its three neighbors. But now you have managed to color the original map with four colors. So the supposed minimal criminal was not criminal after all!

Clearly, no map that aspires to be a minimal criminal can contain a region that borders only three other regions. If it did, you could

(1) shrink that region to a point, thereby reducing the number of regions by one so that the new map falls below the minimal-criminal threshold and *has* to be law-abiding; (2) color the reduced, law-abiding map with four colors; then (3) restore the region you previously shrank by blowing it up again; and (4) color the newly restored region with one of the four colors that doesn't match any of the region's three neighbors, thereby incorporating the restored region into the four-color scheme. So the original map turns out to be law-abiding after all.

A map that can be subjected to this shrink-color-restore process is said to have a "reducible configuration." As we have just seen, one type of reducible configuration is a region that borders only three other regions. Unfortunately, not every map has such a region. But maybe there are other types of reducible configurations. And maybe it could be shown that every map, no matter how complicated, is guaranteed to have at least one reducible configuration. If so, that would clinch the matter. No map that has a reducible configuration can be a minimal criminal; such a map can always be four-colored by the shrink-color-restore process. So if every map has at least one kind of reducible configuration, then there can be no minimal criminals. But no minimal criminals means no criminals, period. (If criminal maps exist at all, some must rank lowest in number of regions.) And no criminals means that every map must be law-abiding—that is, four-colorable.

■

It was by precisely such logic that Alfred Bray Kempe, a London barrister and amateur mathematician, claimed in 1879 to have proved the four-color conjecture. Kempe's reasoning was replete with mathematical convolutions, but it appeared persuasive, and not just to C. S. Peirce. Leading mathematicians in Britain, on the Continent, and in the United States agreed that it was the long-sought solution to the map problem. Kempe was made a fellow of the Royal Society and eventually knighted. His "proof" stood for a decade, until a subtle but fatal flaw was found. The man who detected this flaw was a classicist and mathematician called Percy Heawood—or "Pussy" Heawood by his friends, owing to his catlike whiskers.

Heawood, who was almost apologetic about overturning the result, had personal oddities that make him stand out even in this eccentric-filled saga. A meager, slightly stooping man, he was usually clad in a strangely patterned inverness cape and furnished with an ancient hand-bag. His habitual companion was a dog, who accompanied him to his lectures. He loved to serve on academic committees, and he considered a day "wasted" if it did not include at least one committee meeting. He set his slow-running watch just once a year, on Christmas Day, and then for the following year did the necessary calculations in his head when he needed to know the time. "No, it's not two hours fast," he once insisted to a colleague, "it's ten hours slow!" But he was not with-out practical talents: when the eleventh-century Durham Castle was on the verge of sliding down the cliff on which it was built into the river Wear, Heawood almost single-handedly raised the funds to save it.

And now for another mathematical interlude. If the four-color con-jecture is hard to prove, let's try something easier: what might be called the six-color conjecture. This is like the four-color conjecture but obvi-ously weaker: it says that *six* colors are enough to color any possible map in such a way that neighboring regions are colored differently. The six-color conjecture is worth considering because it reveals the mathemati-cal roots of the map problem, which extend back to the mid-eighteenth century and the great Swiss mathematician Leonhard Euler (whose name is pronounced "oiler"; when I was an undergraduate math stu-dent, we were told to think of the Houston Eulers).

Euler was perhaps the most prolific mathematician in history. Among the discoveries he made, while shuttling between the courts of Freder-ick the Great and Catherine the Great, was the formula $V - E + F = 2$, which was not long ago voted the second most beautiful theorem in mathematics. (The winner of the beauty contest, according to a 1988 survey published in *The Mathematical Intelligencer*, was $e^{i\pi} = -1$.) Euler's formula holds for any polyhedron—that is, for any solid bounded by plane surfaces, such as a cube or a pyramid. It says that if you count up the polyhedron's vertices (V), subtract the number of edges (E), and then add the number of faces (F), the result is always equal to 2. A cube, for instance, has 8 vertices, 12 edges, and 6 faces. And sure enough, $8 - 12 + 6 = 2$.

What do polyhedrons have to do with maps? Well, if you take a polyhedron and (after a little surgical cutting) flatten it out, each face looks like a region of a map. Conversely, you can take a map and knit it up into a polyhedron. The sizes and shapes of the regions will get altered in the process, but that does not affect the overall configuration of the map or the number of colors it needs. The four-color conjecture is thus a problem of topology, the branch of mathematics that is concerned with properties that are not altered by twisting and stretching.

Now, suppose you apply Euler's formula to a map. F becomes the number of regions, E the number of borders, and V the number of points at which borders intersect. Then you can derive a result that is crucial to the map problem: *every map must have at least one region with five or fewer immediate neighbors*. This is agreeably easy to prove. If there were some map in which every region had at least six neighbors, then, by counting up the regions, borders, and points, and plugging them into Euler's formula, you'd get the extraordinary result that $0 = 2$. Contradiction! So every map is guaranteed to have some region that borders on five or fewer other regions.

With this result in hand, it is the work of a moment to prove the six-color conjecture. Suppose there were criminal maps that needed more than six colors. Take a minimal criminal. Now do the old shrink-color-restore trick. Because the supposed minimal criminal must, like all maps, have at least one region with five or fewer neighbors, pick that region and shrink it to a point. Color the reduced map, which must be law-abiding, with six colors. Now reinstate the shrunken region. Since it has five or fewer neighbors—that's why we chose it—there must be a spare color for it among the six available colors. This contradicts the assumption that the map was a minimal criminal, establishing the six-color conjecture as a genuine theorem.

The logic behind all this is a bit circuitous. But if you really try, you can hold it in your mind and "see" why the six-color theorem has to be true. The proof is surprising and yet inevitable at the same time; it's almost witty. Alas, that is as good as the map problem gets, aesthetically speaking. The method that Kempe used in 1879 to get the six-color minimum down to the desired four was tortuous. Yet it drew on no truly deep mathematical ideas. And it contained a fallacious step to boot. Still, when Heawood detected its flaw, he was able to salvage enough of the argument to show that every map could be colored with *five* or fewer colors.

Others attracted to the four-color conjecture included Frederick

Temple, the bishop of London and later archbishop of Canterbury, who also published a fallacious proof; and the French poet Paul Valéry, who left behind a dozen pages of substantial work on the problem in his 1902 diary. Some felt that the nettlesome matter would be dispatched as soon as a really first-rate mathematician turned his attention to it. The great Hermann Minkowski tried to toss off a proof during one of his lectures at the University of Göttingen; after wasting several weeks of classes pursuing it, he announced to his students, "Heaven is angered by my arrogance; my proof is also defective." Other leading mathematicians, perhaps wisely, gave the problem a miss. After all, it was not really in the mainstream of mathematics. Nothing important was known to hang on its truth or falsity. When David Hilbert, perhaps the paramount mathematician of his time, set out before an international conference in Paris in 1900 what he considered the twenty-three most significant problems in mathematics, the four-color conjecture was not among them.

Still, its notoriety and elusiveness continued to make it irresistible to mathematicians on both sides of the Atlantic (some of whom came to regret the time they devoted to it). The ongoing strategy they relied upon was essentially the one Kempe used: find all the loopholes that might permit counterexamples to the four-color conjecture, and then close those loopholes. For this to work, of course, the number of loopholes must be finite; otherwise, they could not all be checked and shown to be closable. As the twentieth century progressed, some mathematicians found ingenious ways of producing complete sets of loopholes; others found equally ingenious ways of closing them.

The problem was that these loophole sets (called unavoidable sets) were absurdly large, consisting of as many as ten thousand map configurations. And closing a given loophole (by showing the configuration in question to be "reducible") could be a monumentally laborious task, one that no human mathematician would be able to carry out. By the 1960s, however, a few people working on the problem had begun to suspect that the loophole-checking process might be made sufficiently routine to be captured by a mechanical algorithm. This raised an interesting possibility: perhaps the four-color conjecture could be proved with the help of a computer.

Mathematicians, it should be noted, were slow to warm to the computer era. Traditionally, from Pythagoras on, they have relied on pure hard thought to gain knowledge of new truths. It used to be said that a mathematics department was the second cheapest for a university to fund, since its members required only pencils, paper, and wastebaskets. (The cheapest would be the philosophers, because they don't need the wastebaskets.) As late as 1986, a mathematician at Stanford boasted that his department had fewer computers than any other, including French literature.

In any case, the four-color conjecture initially seemed too unwieldy even for a computer. It looked as though working through all the cases would take as long as a century on the fastest existing machine. But in the early 1970s, Wolfgang Haken, a mathematician at the University of Illinois, refined the methodology. Together with Kenneth Appel, a gifted programmer, he began a sort of dialogue with the computer, one aimed at reducing the number of loopholes and plugging them more efficiently. "It would work out complex strategies based on all the tricks it had been 'taught,'" Haken later said of the machine, "and often these approaches were far more clever than those we could have tried." Unknown to Appel and Haken, other researchers scattered across the globe, in Ontario and Rhodesia and at Harvard, were closing in on a solution using similar methods. Meanwhile, at least one mathematician was still trying to construct an intricate map that required five colors. In June 1976, after four years of fierce work, twelve hundred hours of computer time, and some crucial help from a professor of literature in Montpellier, France, Haken and Appel got their result: four colors indeed suffice. The story was broken by *The Times* of London that same month. (The warier *New York Times* waited two months to acknowledge the solution, in an op-ed column by the eminent Columbia mathematician Lipman Bers.) The four-color conjecture had become the four-color *theorem*.

Or had it? Whatever the world at large made of this news, many mathematicians reacted sourly when they learned of the details behind it. "Admitting the computer shenanigans of Appel and Haken to the ranks of mathematics would only leave us intellectually unfulfilled," one commented. There were three distinct causes for unhappiness. The first was aesthetic. The proof was unlovely; its bulldozer-like enumeration of cases failed to woo and charm the intellect. And, as G. H. Hardy once declared, "there is no permanent place in the world for ugly mathematics." The second reason had to do with utility. A good proof should contain novel arguments and reveal hidden structures that can be applied elsewhere in mathematics. The Haken-Appel proof seemed sterile in this respect. Nor did it provide any insight into *why* the four-color theorem was true. The answer just sat there like "a monstrous coincidence," as one mathematician put it.

The third and most important reason was epistemological. Did Haken and Appel's accomplishment really furnish grounds for claiming that we *know* the four-color conjecture to be true? Was it really a proof at all? Ideally, a proof is an argument that can be translated into a formal language and verified by the rules of logic. In practice, mathematicians never bother with such formal proofs, which would be cumbersome in the extreme. Instead, they make their arguments reasonably rigorous by spelling out enough of the steps to convince experts in their field. For an argument to be convincing, it must be "surveyable"; that is, it must be capable of being grasped by the human mind and checked for mistakes. And that was certainly not the case with Haken and Appel's proof.

The human part of the argument, comprising some seven hundred pages, was daunting enough. But the *in silico* part, which yielded a four-foot-high computer printout, could never be humanly verified, even if all the mathematicians in the world set themselves to the task. It was as though a key stage in the reasoning had been supplied by an oracle in the form of a long sequence of "yeses." If a single one of these "yeses" should have been a "no," the entire proof would be worthless. How could we be sure that the computer's program did not contain a bug? To check the Haken-Appel result, referees resorted to running a separate computer program of their own, the way a scientist might duplicate an experiment done in a different laboratory. In an influential paper published in 1979 in *The Journal of Philosophy*, "The Four-Color Problem and Its Philosophical Significance," the philosopher Thomas Tymoczko argued that such computer experiments introduced an empirical element into mathematics. Almost all mathematicians now believe that the four-color theorem is true, but they do so on the basis of corrigible evidence. The theorem conspicuously fails to conform to the Platonic ideal of certain, absolute, a priori knowledge.

The most we can say is that it is probably true, like the theories of physics that underwrite the operation of the machine that helped establish it.

The four-color breakthrough marked a shift in mathematical practice. Since then, several other conjectures have been resolved with the aid of computers (notably, in 1988, the nonexistence of a projective plane of order 10). Meanwhile, mathematicians have tidied up the Haken-Appel argument so that the computer part is much shorter, and some still hope that a traditional, elegant, and illuminating proof of the four-color theorem will someday be found. It was the desire for illumination, after all, that motivated so many to work on the problem, even to devote their lives to it, during its long history. (One mathematician had his bride color maps on their honeymoon.) Even if the four-color theorem is itself mathematically otiose, a lot of useful mathematics got created in failed attempts to prove it, and it has certainly made grist for philosophers in the last few decades. As for its having wider repercussions, I'm not so sure. When I looked at the map of the United States in the back of a huge dictionary that I once won in a spelling bee for New York journalists, I noticed with mild surprise that it was colored with precisely four colors. Sadly, though, the states of Arkansas and Louisiana, which share a border, were both blue.

PART V

Infinity, Large and Small

11

Infinite Visions: Georg Cantor v. David Foster Wallace

Few ideas have had a racier history than the idea of infinity. It arose amid ancient paradoxes, proceeded to baffle philosophers for a couple of millennia, and then, by a daring feat of intellect, was finally made to yield its secrets in the late nineteenth century, though not without leaving a new batch of paradoxes. You don't need any specialized knowledge to follow the plot: the main discoveries, despite the ingenuity behind them, can be conveyed with a few strokes of a pen on a cocktail napkin. All of this makes infinity irresistible meat for the popularizer, and quite a few books in that vein have appeared over the years. The most extraordinary figure to try his hand at this was David Foster Wallace. As readers of *Infinite Jest* might suspect, its author had a deep and sophisticated grasp of mathematics and metaphysics. *Everything and More: A Compact History of* ∞—written five years before Wallace's suicide in 2008 at the age of forty-six—was his attempt to initiate the mathematically lay reader into the mysteries of the infinite.

It might seem odd that finite beings like us could come to know anything about infinity, given that we have no direct experience of it. Descartes thought that the idea of infinity was innate, but the behavior of children suggests otherwise; in one study, children in their early school years "reported 'counting and counting' in an attempt to find the last number, concluding there was none after much effort." As it happens, the man who did the most to capture infinity in a theory claimed that his insights were vouchsafed to him by God and ended his life in a mental asylum.

Broadly speaking, there are two versions of infinity. The woollier, more mystical one, which might be called metaphysical infinity, is associated with ideas like perfection, the absolute, and God. The more hardheaded version, mathematical infinity, is the one that Wallace set out to explicate. It derives from the idea of endlessness: numbers that can be generated inexhaustibly, time that goes on forever, space that can be subdivided without limit. While metaphysical infinity tends to evoke awe in those who contemplate it, mathematical infinity has, for most of Western intellectual history, been an object of grave suspicion, even scorn. It first cropped up in the fifth century B.C.E., in the paradoxes of Zeno of Elea. If space is infinitely divisible, Zeno argued, then swift Achilles could never overtake the tortoise: each time he caught up to where the tortoise was, it would have advanced a little farther, ad infinitum. So traumatizing were such paradoxes that Aristotle was moved to ban the idea of a "completed" infinity from Greek thought, setting the orthodoxy for the next two thousand years.

The eventual rehabilitation of the infinite had its origins in another paradox, this one framed by Galileo in 1638. Consider, Galileo said, all the whole numbers: 1, 2, 3, 4, and so on. Now consider just those numbers that are squares: 1, 4, 9, 16, and so on. Surely there are more numbers than squares, because the squares make up just a part of the numbers, and a small part at that. Yet, Galileo observed, there is a way of pairing off squares with whole numbers: 1 to 1, 2 to 4, 3 to 9, 4 to 16, and so on. When two finite sets can be made to correspond in this way, so that each item from the first set gets matched with precisely one item from the second set, and vice versa, you know that they are of equal size without having to go through the tedious business of counting. Extending the principle to infinite collections, Galileo felt himself drawn toward the conclusion that there are as many square numbers as there are numbers, period. The part, in other words, was equal to the whole—a result that struck him as absurd.

Two and a half centuries later, Georg Cantor made Galileo's paradox the basis for a mathematical theory of infinity. Cantor, who lived from 1845 to 1918, was a Russian-born German mathematician with an artistic streak and a keen interest in theology. He saw that the collapse of the familiar logic of part-and-whole yielded a new definition of infinity, one that did not rely on the vague notion of endlessness. An infinite set, as Cantor characterized it, is one that is the same size as some of its parts. In other words, an infinite set is one that can lose some of its members without being diminished.

Now Cantor was in a position to ask a novel question: Are all infinities the same, or are some more infinite than others?

Looking for an infinity bigger than that of the whole numbers, Cantor started by considering the set of fractions. It seemed a good bet, because fractions are densely ordered on the number line: between every two whole numbers, there are infinitely many fractions. (Between 0 and 1, for example, lie $1/2$, $1/3$, $1/4$, $1/5$, and so on.) Yet Cantor, to his surprise, was able to find an easy way of matching up the whole numbers and the fractions one to one. Despite initial appearances, these two infinities turned out to be the same. Perhaps, he thought, all infinite sets, by dint of being inexhaustible, were of the same magnitude. But then he looked at the "real" numbers—those marking off the points on a continuous line. Could they, too, be paired up, one to one, with the whole numbers? By a surpassingly clever bit of reasoning, called the diagonal proof, Cantor showed that the answer was no. In other words, there were at least two distinct infinities, that of the whole numbers and that of the continuum, and the second was greater than the first.

But was that the end of it? In searching for a still-larger species of infinity, Cantor looked to higher dimensions. Surely, he thought, there must be more points in a two-dimensional plane than on a one-dimensional line. For a couple of years, he struggled to prove that the points in a plane could not be paired off one to one with the points in a line, only to find, in 1878, that such a correspondence was indeed possible. A simple trick showed that there were exactly as many points on a one-inch line as there were in all of space. "I see it, but I do not believe it!" Cantor wrote to a colleague.

With the discovery that neither size nor dimension was the way to higher infinities, the quest seemed to stall. But after more than a decade of intense work (interrupted by a stay in a sanatorium after a nervous breakdown), Cantor arrived at a powerful new principle that enabled him to resume the ascent: there are always more *sets of* things than things. This is obvious enough in a finite world. If you have, say, three objects, you can form eight different sets out of them (including, of course, the empty set). Cantor's genius lay in extending the same principle into the realm of the infinite.

To make matters a little less abstract, you might pretend that we live in a world with an infinite number of people. Now consider all the possible clubs (sets of people) that might exist in this world. The *least* exclusive of these clubs—the universal club—will be the one of which absolutely everyone is a member. The *most* exclusive—the null club—will be the one that has no members at all. In between these two extremes will lie an infinity of other clubs, some with lots of members, some with few. How big is this infinity? Is there any way of matching up clubs and people one to one, thereby showing that the two infinite collections are the same size? Suppose that every person can be paired off with precisely one club, and vice versa. Some of the people will happen to be members of the club with which they are paired (for example, the person who gets paired with the universal club). Others will happen *not to be* members of their associated club (for example, the person who gets paired with the null club). Those people make up a group you might call the Groucho Club. The Groucho Club is a sort of *salon des refusés*: it consists of all the people who are paired with clubs that won't have them as members. So the person who gets paired with the null club—which, of course, excludes him—at least has the consolation of being a member of the Groucho Club.

Now, here is where things get interesting. Since the pairing of people with clubs is assumed to be complete, there must be some fellow who is paired up with the Groucho Club itself. Call him Woody. Is Woody a member of the Groucho Club or not? Well, suppose he is. That means that by definition he must be excluded from the club he is paired with. So Woody is *not* a member of the Groucho Club. But if he is not a member of the Groucho Club, then, because the club he is paired with won't have him, he *is* a member of the Groucho Club. No matter which way we turn, we get a contradiction. How did we get to this impasse? By supposing that people could be paired off, one to one, with clubs. So that supposition has to be false. Which is to say, the infinity of sets of things is greater than the infinity of things.

The beauty of this principle, which has come to be known as Cantor's theorem, is that it can be applied over and over again. Given any infinite set, you can always come up with a larger infinity by considering its "power set"—the set of all subsets that can be formed from it. Atop a simple reductio ad absurdum, Cantor built a never-ending tower of infinities. It seemed a dream vision, like Coleridge's "Kubla Khan." Yet mathematicians found in this new theory the resources needed to put their subject on a sure footing. "No one shall expel us from the paradise which Cantor has created for us," the great (and influential) mathematician David Hilbert declared. Others, however, dismissed Cantor's infinity of infinities as a "fog on a fog" and "mathematical insanity." Cantor felt persecuted by these critics, which worsened his nervous condition (he seems to have suffered from a bipolar disorder). Between frequent breakdowns and hospitalizations, he pondered the theological implications of the infinite and, with equal ardor, pursued the theory that Bacon wrote the works of Shakespeare.

Cantor's theory constitutes "direct evidence that actually-infinite sets can be understood and manipulated, truly *handled* by the human intellect," Wallace wrote in *Everything and More*. What makes this achievement so heroic, he observed, is the awful abstractness of infinity: "It's sort of the ultimate in drawing away from actual experience," a negation of "the single most ubiquitous and oppressive feature of the concrete world—namely that everything ends, is limited, passes away." Wallace was alive to the "dreads and dangers" of abstract thinking. For two millennia, the idea of infinity was thought to pose hazards to one's sanity. It was Cantor, for all his madness, who managed to tame infinity and showed that it could be reasoned about without derangement.

Writing accessibly about abstract mathematical ideas poses hazards of its own. One pitfall of such attempts is purple prose. A widely read book about the calculus tells us that "the Cartesian plane is suffused with a strange and somber silence"; a book about zero says that this number is "a shadow in the slanting light of fear." Another pitfall is mysticism. In *The Mystery of the Aleph*, written a few years before Wallace's own book on infinity, the mathematician Amir D. Aczel tries to make Georg Cantor into a Kabbalist who entered "God's secret garden" and lost his sanity in punishment. Rudy Rucker's *Infinity and the Mind*, a terrific study with real mathematical depth, takes an unwelcome excursion into Zen Buddhism. On the other hand, a little classic called *Playing with Infinity*, by Rózsa Péter, a Hungarian logician who died in 1977, achieves both charm and clarity without a bit of extraneous guff. But all these popularizers, whatever their lapses, conscientiously did the brutal work that is required to make abstract ideas clear, even beautiful, to the beginner. By simplifying, leaving things out, they produced a first approximation to real understanding.

Wallace's effort, by contrast, can't quite be described as popularization. Wallace assured the reader that it is "a piece of pop technical writing" and claimed that his own math background didn't go much beyond high school. And yet he refused to make the usual compromises. *Everything and More* is sometimes as dense as a math textbook, though rather more chaotic. I have never come across a popular book about infinity that packs so much technical detail—especially one that purports to be "compact." (What Wallace calls a "booklet" actually runs to more than three hundred pages; however, his book is "compact" in the mathematician's technical sense, because it is bounded and it can be closed.) Wallace's motive was admirable: he was determined to improve on "certain recent pop books that give such shallow and reductive accounts of Cantor's proofs . . . that the math is distorted and its beauty obscured." But when a writer's grasp of his material is less than sure, he risks sacrificing clarity to spectacle, flashing equations and technical terms at his audience like a magician's deck of cards. Wallace invited readers—presumably newcomers to the mathematics of infinity—simply to "revel" in the "symbology." He told them that some of the terms—"inaccessible ordinals," "transfinite recursion"—are "fun to say even if one has no clear idea what they're supposed to denote." An aesthetic fondness for the math textbook's visual display may also explain his penchant for initials and abbreviations ("with respect to" becomes "w/r/t," Galileo becomes "G.G.," and the "Divine Brotherhood of Pythagoras" is "D.B.P.")—a penchant that was also frequently indulged in his works of fiction.

Still, Wallace's enthusiasm for the theory of infinity is evident on every page (not least in his conviction that Cantor is "the most important mathematician of the nineteenth century," a view that few mathematicians or intellectual historians would agree with). And if he was sometimes in over his head, it is because he chose to wade through the deepest waters. The question is whether he can bring his readers along.

A book that prizes difficulty but not rigor is probably not meant for those in search of mathematical illumination. What Wallace offered, in the end, *is* a purely literary experience. Regarding the nature of that experience, you may find a clue in Ludwig Wittgenstein's response to Cantor's great contribution. The thrill we get from discovering that some infinities are bigger than others, Wittgenstein thought, is just a "schoolboy pleasure." There is nothing awesome about the theory; it does not describe a world of timeless, transcendent, scarcely conceivable entities; it is really no more than a collection of (finite) tricks of reasoning. One might imagine, Wittgenstein said, that the theory of infinite sets was "created by a satirist as a kind of parody of mathematics." Wallace, whose satiric gifts were transcendent, might have achieved something considerable after all—a sly send-up of pop technical writing. "A parody of mathematics": As a description of Cantor's work on infinity, it is surely unjust. As a description of Wallace's, it may be taken as a tribute.

12

Worshipping Infinity: Why the Russians Do and the French Don't

Mathematics and mysticism go way back together. Higher mathematics was invented by the Pythagoreans, a cult whose tenets included the transmigration of souls and the sinfulness of eating beans. Even today, there remains a whiff of the mystical about mathematics. Many mathematicians, even quite prominent ones, openly avow their belief in a realm of perfect mathematical entities hovering over the grubby empirical world—a sort of Platonic heaven.

One such Platonist is Alain Connes, who holds the chair in analysis and geometry at the Collège de France. In a dialogue a couple of decades ago with the neurobiologist Jean-Pierre Changeux, Connes declared his conviction that "there exists, independently of the human mind, a raw and immutable mathematical reality," one that is "far more permanent than the physical reality that surrounds us." Another unabashed Platonist is Sir Roger Penrose, emeritus Rouse Ball professor of mathematics at Oxford, who holds that the natural world is but a "shadow" of a Platonic realm of eternal mathematical forms.

The rationale for this otherworldly view of mathematics was first framed by Plato himself, in the *Republic*. Geometers, he observed, talk of perfectly round circles and perfectly straight lines, neither of which are to be found in the sensible world. The same is true even of numbers, he held, since they must be composed of perfectly equal units. Plato concluded that the objects studied by mathematicians must exist in another world, one that is changeless and transcendent.

Seductive though the Platonic picture of mathematics might be, it leaves one thing mysterious. How are mathematicians supposed to get in touch with this transcendent realm? How do they come to have knowledge of mathematical objects if those objects lie beyond the world of space and time? Contemporary Platonists tend to do a bit of hand waving when confronted with this question. Connes invokes a "special sense," one "irreducible to sight, hearing or touch," that allows him to perceive mathematical reality. Penrose avers that human consciousness somehow "breaks through" to the Platonic world. Kurt Gödel, among the staunchest of twentieth-century Platonists, wrote that "despite their remoteness from sense experience, we do have something like a perception" of mathematical objects, adding, "I don't see any reason why we should have less confidence in this kind of perception, i.e., in mathematical intuition, than in sense perception."

But mathematicians, like the rest of us, think with their brains. And it is hard to see how a physical system like the brain could interact with a nonphysical reality. As the philosopher Hilary Putnam observed, "We cannot envisage *any* kind of neural process that could even correspond to the 'perception of a mathematical object.'"

One way out of this dilemma is to throw over Plato for Aristotle. There may be no perfect mathematical entities in our world, but there are plenty of imperfect approximations. We can draw crude circles and lines on a chalkboard; we can add two apples to three apples, even if these apples are not identical, and end up with five apples. By abstracting from such experience of ordinary perceptible things, we arrive at basic mathematical intuitions. And logical deduction does the rest.

This is the Aristotelian view of mathematics, and it pretty much accords with common sense. But there is one putative mathematical object that it can't handle: infinity. We have no experience of the infinite. We have no experience of anything *like* it. True, we do have a sense of numbers going on indefinitely—because, no matter how big a number we think of, we can always get a bigger number by adding 1 to it. And we think we can imagine space or time extending without limit. But an actual "completed" infinity, as opposed to a merely "potential one"— this is something we never encounter in the natural world.

The idea of infinity was long regarded with suspicion, if not horror. Zeno's paradoxes seemed to show that if space could be divided up infinitely into infinitesimal segments, then motion would be impossible. Aquinas argued that infinite numbers were inherently contradictory, since numbers arise from counting and an unlimited collection could never be counted. Galileo observed that infinity appeared to violate the principle that the part must be less than the whole. Speculation about the infinite was left to theologians, who identified it with the divine. For Pascal, whose seventy-second *pensée* is a prose poem to infinity, the infinite is something "not to understand, but to admire." As late as 1831, Gauss declared that "the use of an infinite quantity as an actual entity . . . is never allowed in mathematics."

But it became apparent that mathematicians could not do without infinity. Even the "applied" part of mathematics—the mathematical physics that grew out of Newton's and Leibniz's invention of the calculus—had persistent glitches at its foundation, glitches that only a rigorous theory of sets, including infinite sets, could fix. It was in the late nineteenth century that Georg Cantor, a Russian-born German mathematician, supplied the needed theory. Cantor did not set out to characterize the infinite for its own sake; rather, he claimed, the idea "was logically forced upon me, almost against my will."

What led Cantor to develop his theory of infinite sets was the homely-sounding problem of the "vibrating string." What he ended up with, after a couple of decades of intellectual struggle, was anything but homely: a succession of higher and higher infinities—an infinite hierarchy of them, ascending toward an unknowable terminus he called the Absolute. This seemed to him a divinely vouchsafed vision; in transmitting it to the world, he regarded himself (in the words of his biographer Joseph Dauben) as "God's ambassador."

Cantor's new theory initially got a mixed reception. His onetime teacher Leopold Kronecker reviled it as "humbug" and "mathematical insanity," whereas David Hilbert declared, "No one shall expel us from the paradise which Cantor has created for us." Bertrand Russell recalled in his autobiography that he "falsely supposed all [Cantor's] arguments to be fallacious," only later realizing that "all the fallacies were mine."

In some cases, the reaction to Cantor's theory broke along national lines. French mathematicians, on the whole, were wary of its metaphysical aura. Henri Poincaré (who rivaled Germany's Hilbert as the greatest mathematician of the era) observed that higher infinities "have a whiff of form without matter, which is repugnant to the French spirit." Russian mathematicians, by contrast, enthusiastically embraced the newly revealed hierarchy of infinities.

Why the contrary French and Russian reactions? Some observers have chalked it up to French rationalism versus Russian mysticism. That is the explanation proffered, for example, by Loren Graham, an American historian of science retired from MIT, and Jean-Michel Kantor, a mathematician at the Institut de Mathématiques de Jussieu in Paris, in their book *Naming Infinity* (2009). And it was the Russian mystics who better served the cause of mathematical progress—so argue Graham and Kantor. The intellectual milieu of the French mathematicians, they observe, was dominated by Descartes, for whom clarity and distinctness were warrants of truth, and by Auguste Comte, who insisted that science be purged of metaphysical speculation. Cantor's vision of a never-ending hierarchy of infinities seemed to offend against both.

The Russians, by contrast, warmed to the spiritual nimbus of Cantor's theory. In fact, the founding figures of the most influential school of twentieth-century Russian mathematics were adepts of a heretical religious sect called the Name Worshippers. Members of the sect believed that by repetitively chanting God's name, they could achieve fusion with the divine. Name Worshipping, traceable to fourth-century Christian hermits in the deserts of Palestine, was revived in the modern era by a Russian monk called Ilarion. In 1907, Ilarion published *On the Mountains of the Caucasus*, a book that described the ecstatic experiences he induced in himself while chanting the names of Christ and God over and over again until his breathing and heartbeat were in tune with the words.

To the Russian Orthodox episcopacy, the Name Worshippers were guilty of the heresy of equating God with his name, and the tsarist regime acted to suppress the movement (in one case sending in Russian marines to storm a monastery atop Mount Athos on the Aegean, full of rebellious Name-Worshipping monks). But for the sect's mathematical adherents, Name Worshipping seemed to open up a special avenue to the infinite and to the Platonic heaven where it lived. The Russians were thus emboldened to make free use of higher infinities in their mathematical work. "While the French were constrained by their rationalism, the Russians were energized by their mystical faith," Graham and Kantor declare.

Two distinct questions are raised here. First, did Name-Worshipping mysticism really give the Russians a leg up mathematically? Graham and Kantor are pretty sure it did, claiming that in this case "a religious heresy was instrumental in helping the birth of a new field of modern mathematics." That leads to a second question: Can mysticism play a genuine role in the attainment of mathematical knowledge, especially knowledge of the infinite? Here the authors, confirmed secularists as they are, are less certain. "We trust rational thought more than mystical inspiration," they say. The same, though, could be said of the French mathematicians who were supposedly surpassed by the Russians. One is left with the impression that mysticism in mathematics has at least a degree of pragmatic truth; that is, it works.

Consider the conceptual challenges that mathematicians were facing toward the end of the nineteenth century. When Cantor began his work on infinity, the fundamental concepts of calculus—long the most important branch of mathematics for understanding the physical world—were still in a muddle. Essentially, calculus deals with curves. Its two basic operations are finding the direction of a curve at a given point (the "derivative") and the area bounded by a curve (the "integral"). Curves are mathematically represented by "functions." Some functions, like the sine wave, are nice and smooth; they are called continuous. But others are riddled with breaks and jumps: discontinuities. Just how discontinuous could a function be and still be handled by the methods of calculus? That was a critical question with which Cantor's contemporaries were struggling.

The key to answering it proved to be the idea of a *set*. Consider the collection of all the points where a function takes discontinuous jumps. The larger and more complicated this set of discontinuities is, the more "pathological" the function. So Cantor's attention was drawn to sets of points. How could the size of such a set be measured? In answering this question, Cantor was led to a theory in which an entire hierarchy of infinities, distinguished by their size, can be defined.*

Cantor's theory of sets, and his distinction between "small" and "large" infinities, supplied what was needed to shore up the calculus

---

*For the details of Cantor's theory, see pp. 132–35 in this volume.

and extend its basic concepts. A trio of French mathematicians led the way. Émile Borel, a mathematician who directed the École Normale Supérieure, was also a journalist (he published the influential leftist organ *Revue du Mois*), a government minister, a fixture in the Parisian *beau monde*, and, eventually, a *résistant* and prisoner of the Gestapo. He and his students Henri Lebesgue and René Baire resolved some of the most vexing issues in the foundations of calculus. Borel launched what became known as measure theory, which became the basis of the study of probability. Baire deepened the understanding of continuity and its relation to the derivative. And Lebesgue produced a beautiful new theory of the integral that eliminated its most annoying flaws.

These magnificent achievements all built on Cantor's work, yet the French trio had reservations about it. Paradoxes discovered by Bertrand Russell and others made them worried that the new set theory might be logically unsound. They were especially skeptical of a novel assumption called the axiom of choice, which was introduced in 1904 by the German mathematician Ernst Zermelo to extend Cantor's theory. The axiom of choice asserts the existence of certain sets even when there is no recipe for producing them. Suppose, for example, you start with a set consisting of an infinite number of pairs of socks. Let's say you want to define a new set consisting of just one sock from each pair. Since the socks in a pair are identical, there is no rule for doing this. The axiom of choice nevertheless guarantees that such a set exists, even though it represents an infinite number of arbitrary choices.

The French trio ultimately rejected the axiom of choice—"such reasoning does not belong to mathematics," Borel declared—and, with it, the use of higher infinities. Was this intellectual timidity? The authors of *Naming Infinity* think so. The French trio "lost their nerve"; they "confronted an intellectual abyss before which they came to a halt." And the price they paid for their qualms, we are given to believe, was psychological as well as mathematical. Borel retreated from the abstractions of set theory to the safer ground of probability theory. "Je vais pantoufler dans les probabilités," as he charmingly put it ("I'm going to dally with probability"; *pantoufler* literally means "play around in my slippers"). Lebesgue in his "frustration" became "somewhat sour." Baire, whose physical and mental health were always delicate, ended his life in solitude and suicide.

Their Russian trio of counterparts, by contrast, welcomed the metaphysical aspects of set theory. The senior figure of the Russian trio, Dmitri Egorov, was a deeply religious man. So was his student Pavel Florensky, a mathematician who also trained as a priest. (Some years into the Bolshevik era, the sight of Father Florensky addressing a scientific conference in his clerical robes moved an incredulous Trotsky to exclaim, "Who is *that*?") Florensky became the spiritual mentor of another of Egorov's students, Nikolai Luzin. Both Egorov and Florensky were members of an underground cell of the Name-Worshipping sect, which had spread from rural monasteries to the Moscow intelligentsia, and Luzin, if not an actual member, was influenced by its philosophy.

All three of these Russians carried their Name Worshipping over into mathematics. They seemed to believe that the very act of naming could put them in touch with infinite sets undefinable by ordinary mathematical means. "Can we convince ourselves of the existence of a mathematical object without defining it?" Lebesgue had asked skeptically. To Florensky, this was like asking, "Is it possible to convince oneself of the existence of God without defining him?" Of course it was, the Russians held; the very name of God, when repeatedly invoked, carried the conviction of his existence. (Indeed, the unofficial slogan of the Name Worshippers was "The name of God *is* God.") The Russians were convinced they could summon new mathematical entities into existence merely by naming them.

It is hard to imagine how a name might possess such magical powers. In contemporary philosophy, there are two rival theories of how names work. According to the "descriptivist" theory (which originated with the German logician Gottlob Frege), a name has an associated description, and the thing the name refers to is what satisfies that description. When we use the name Homer, for instance, we are referring to whatever individual satisfies a description like "author of the *Iliad* and the *Odyssey*." The more recent, "causal" theory of names (defended by the American philosopher Saul Kripke, among others) denies that names have an associated descriptive sense; instead, it says, a name is attached to its bearer by a historical chain of communication reaching in space and time all the way back to an initial act of baptism. According to one theory, names are stuck onto their bearers by semantic glue; according to the other, by causal glue.

Which of these theories works for naming mathematical objects? Not the causal theory, evidently. There is no way for a mathematician to get into causal contact with infinity. You can't point your finger at an infinite set and say, "I dub thee A," because such sets, if they exist, are not part of the spatiotemporal world. The only way to name an infinite set is by producing a mathematical description that it uniquely satisfies, as presupposed by the descriptivist theory of names. So you might name a certain infinite set by saying, "Let A be the set of all rational numbers whose squares are less than 2." Here, of course, the name is a mere shorthand convenience. What actually does the referential work is the definition. And without that definition, there is no way to assert that set's existence.

That is what the French trio realized. "To define always means naming a characteristic property of what is being defined," Lebesgue wrote. *Defining* a thing means citing a property that distinguishes it from other things. And this is just the sort of definition that the axiom of choice allowed one to dispense with—dangerously, in the French view.

Did the freewheeling use of infinity by the mystical Russians really enable them to make breakthroughs denied to the more cautious French? The authors of *Naming Infinity* claim as much, but they exaggerate a bit. It was the French trio, after all, who wrote the dramatic final chapter in the logical development of calculus. Every working mathematician is intimately acquainted with the "Borel algebra," the "Baire category theorem," and above all the "Lebesgue integral." To this chapter the Russian trio merely added a few footnotes. (Egorov's most famous theorem, which concerns an infinite sequence of functions, was essentially a rediscovery of a result due to Borel and Lebesgue.) It is true that Luzin helped found "descriptive set theory," a subbranch of set theory that uses Cantor's higher infinities to describe complicated subsets of the real-number line. But to call this "a new field of modern mathematics," as the authors of *Naming Infinity* do, is greatly to inflate its importance.

The real achievement of Egorov and Luzin was to put Moscow on the mathematical map. A circle of young mathematicians at Moscow University formed around them in the early 1920s, taking the name Lusitania in Luzin's honor. "The Lord himself—professor Luzin— / Shows us the pathway to research!" one Lusitanian rhapsodized in a poem. A florescence of mathematical creativity occurred amid famine and civil war. Seminars were held in subfreezing temperatures because of fuel shortages, but the students coped by creating an ice-skating rink inside the mathematics building, "singing while they glided on the ice around the central staircase under the skylight."

In the early years of the Soviet era, mathematicians were largely left alone by the authorities because of the abstract nature of their work. Egorov and Luzin left religion out of their lectures, hinting only at the "mystical beauty" of the mathematical world and the importance of bestowing names on its objects. But the relative permissiveness ended when Stalin came to power. Egorov was denounced as "a reactionary supporter of religious beliefs, a dangerous influence on students, and a person who mixes mathematics and mysticism." His accuser was Ernst Kol'man, an impish and sinister Marxist mathematician nicknamed the "dark angel." Egorov and Florensky, along with other Name Worshippers, were eventually arrested. Egorov starved to death in prison in 1931, his last words reputedly being "Save me, O God, by Thy name!" Florensky was tortured and sent to a Gulag camp in the Arctic, where he was probably executed in 1937. Luzin, too, was targeted by Kol'man, who, in a Monty Python–ish touch, deployed esoteric mathematical arguments against him. But Luzin had powerful defenders, one of whom, appealing to Stalin, observed that Newton himself had been a "religious maniac." After undergoing a humiliating trial—where, among other charges, he was accused of publishing his results in foreign journals—he was spared.

Several of Luzin's former students took part in the campaign against him, among them Pavel Alexandrov and Andrei Kolmogorov. Both of these students ended up outshining their mentor, and Kolmogorov is now rated one of the half a dozen greatest mathematicians of the twentieth century. Kolmogorov and Alexandrov became longtime gay lovers; their favorite activity was swimming vast distances and then doing mathematics together in the nude. The hostility they showed Luzin was probably more a matter of professional rivalry and personal friction than ideology. On one occasion, Luzin insulted Kolmogorov on the floor of the Russian Academy of Sciences by making an elaborate (and untranslatable) pun about sodomy and higher mathematics, whereupon Kolmogorov struck him in the face.

The Moscow school of mathematics flourished long after the eclipse of its mystically inclined founders. In the postwar era, only Paris rivaled the Russian capital as a center of mathematical talent. But the higher infinities revered by the Name Worshippers ceased to be of great importance, and the elaborate set theory favored by Luzin got displaced by the more mainstream methods of Kolmogorov and Alexandrov. As for the once controversial axiom of choice, Kurt Gödel removed any need for a mystical rationale when he proved in 1938 that it was logically consistent with the other, generally accepted axioms of set theory. Since no harmful contradictions could result from its use, mathematicians were now free to employ the axiom as they saw fit. They did not have to worry about whether it truly described a Platonic world of infinite sets.

Herein lies a clue to the demystification of mathematics. What gives the subject a mystical tincture is the reputed nature of its objects. Whereas the objects studied by botany or chemistry are part of the physical world, those of mathematics are thought to inhabit a transcendent world that cannot be accessed by normal modes of knowledge. But suppose there are no such transcendent objects. Does mathematics then become like theology without God? Is it (as philosophical nominalists insist) mere make-believe, a marvelously complicated fairy tale?

In a sense, yes. If there is no actual mathematical reality to be described, mathematicians are free to make up stories—that is, to explore whatever hypothetical realities they can imagine. As Cantor himself once declared, "The essence of mathematics is freedom." On this picture, their work consists of if-then assertions: *if* such-and-such structure satisfies certain axioms, *then* that structure must satisfy certain further conditions. (This if-then view of mathematics has been held at times by Bertrand Russell and Hilary Putnam, among others.) Some of these axioms might describe hypothetical structures that have analogues in the physical world, making for useful "applied" mathematics. Others may be irrelevant to understanding the physical world but still have utility within mathematics. The axiom of choice, for example, is not needed for applied mathematics, but most mathematicians choose to employ it because it streamlines the more make-believe areas of mathematics, like topology.

There is only one constraint on the storytelling of mathematicians (other than the need to get tenure), and that is consistency. As long as a collection of axioms is consistent, then it describes some possible structure. But if the axioms turn out to be inconsistent—that is, if they harbor a contradiction—they can describe no possible structure and are hence a waste of the mathematician's time.

Mathematics as a style of reasoning rather than a science of transcendent objects: Is that too disenchanted a picture to sustain the working mathematician? "In the history of mathematics, from the time of Pythagoras to the present, there have been periods of waxing and waning of the elements of rationalism and mysticism," the authors of *Naming Infinity* observe. Today, the romance of a Platonic mathematical reality is still very much alive, as evidenced by the remarks of Alain Connes and Roger Penrose quoted earlier.

And then there is the more baroque case of Alexander Grothendieck. Working in Paris in the 1960s, Grothendieck (the son of a Russian anarchist who died at Auschwitz) created a new abstract framework that revolutionized mathematics, enabling its practitioners to express hitherto inexpressible ideas. Grothendieck's approach had a strong mystical bent. In his voluminous autobiographical writings, he described a creative process involving "visions" and "messenger dreams." Like the Russian Name Worshippers, the authors of *Naming Infinity* note, he saw naming "as a way to grasp objects even before they have been understood."

Grothendieck might be held up as an advertisement for the pragmatic power of mysticism in mathematics. He died in 2014, a recluse in the Pyrenees—where, according to the rare visitor, he spent the last decades of his life "obsessed with the Devil, which he sees at work everywhere in the world, destroying the divine harmony."

13

The Dangerous Idea of the Infinitesimal

When people talk about the infinite, they usually mean the infinitely great: inconceivable vastness, world without end, boundless power, the absolute. There is, however, another kind of infinity that is quite different from these, though just as marvelous in its own way. That is the infinitely small, or the infinitesimal.

In everyday parlance, "infinitesimal" is loosely used to refer to things that are extremely tiny by human standards, too small to be worth measuring. It tends to be a term of contempt. In his biography of Frederick the Great, Carlyle tells us that when Leibniz offered to explain the infinitely small to Queen Sophia Charlotte of Prussia, she replied that on that subject she needed no instruction: the behavior of her courtiers made her all too familiar with it. About the only non-pejorative use of "infinitesimal" I have come across occurs in Truman Capote's unfinished novel *Answered Prayers*, when the narrator is talking about the exquisite vegetables served at the tables of the really rich: "the greenest *petits pois*, infinitesimal carrots." Then there are the abundant malapropisms. Some years back, *The New Yorker* reprinted a bit from an interview with a Hollywood starlet in which she was describing how she took advantage of filming delays on the set to balance her checkbook, catch up on her mail, and so forth. "If you really organize your time," she observed, "it's almost infinitesimal what you can accomplish." (To which *The New Yorker* ruefully added, "We know.")

Properly speaking, the infinitesimal is every bit as remote from us as the infinitely great is. Pascal, in his seventy-second *pensée*, pictured nature's "double infinity" as a pair of abysses between which finite man is poised. The infinitely great lies without, at the circumference of all things; the infinitesimal lies within, at the center of all things. These two extremes "touch and join by going in opposite directions, and they meet in God and God alone." The infinitely small is even more difficult for us to comprehend than the infinitely great, Pascal observed: "Philosophers have much oftener claimed to have reached it, [but] they have all stumbled."

Nor, one might add, has the poetical imagination been much help. There have been many attempts in literature to envisage the infinitely great: Father Arnall's sermon on eternity in *A Portrait of the Artist as a Young Man*, Borges's quasi-infinite "Library of Babel." For the infinitesimal, though, there is only vague talk from Blake about an infinity you can hold "in the palm of your hand," or, perhaps more helpful, these lines from Swift: "So, naturalists observe, a flea / Hath smaller fleas that on him prey; / And these have smaller still to bite 'em, / And so proceed *ad infinitum*."

From the time it was conceived, the idea of the infinitely small has been regarded with deep misgiving, even more so than that of the infinitely great. How can something be smaller than any given finite thing and not be simply nothing at all? Aristotle tried to ban the notion of the infinitesimal on the grounds that it was an absurdity. David Hume declared it to be more shocking to common sense than any priestly dogma. Bertrand Russell scouted it as "unnecessary, erroneous, and self-contradictory."

Yet for all the bashing it has endured, the infinitesimal has proved itself the most powerful device ever deployed in the discovery of physical truth, the key to the scientific revolution that ushered in the Enlightenment. And, in one of the more bizarre twists in the history of ideas, the infinitesimal—after being stuffed into the oubliette seemingly for good at the end of the nineteenth century—was decisively rehabilitated in the 1960s. It now stands as the epitome of a philosophical conundrum fully resolved. Only one question about it remains open: Is it real?

■

Ironically, it was to save the natural world from unreality that the infinitesimal was invoked in the first place. The idea seems to have appeared in Greek thought sometime in the fifth century B.C.E., surfacing in the great metaphysical debate over the nature of being. On one side of this debate stood the monists—Parmenides and his followers— who argued that being was indivisible and that all change was illusion. On the other stood the pluralists—including Democritus and his fellow atomists, as well as the Pythagoreans—who upheld the genuineness of change, which they understood as a rearrangement of the parts of reality.

But when you start parsing reality, breaking up the One into the Many, where do you stop? Democritus held that matter could be analyzed into tiny units—"atoms"—that, though finite in size, could not be further cut up. But space, the theater of change, was another question. There seemed to be no reason why the process of dividing it up into smaller and smaller bits could not be carried on forever. Therefore, its ultimate parts must be smaller than any finite size.

This conclusion got the pluralists into a terrible bind, thanks to Parmenides's cleverest disciple, Zeno of Elea. Irritated (according to Plato) by those who ridiculed his master, Zeno composed no fewer than forty dialectical proofs of the oneness and changelessness of reality. The most famous of these are his four paradoxes of motion, two of which—the "dichotomy" and "Achilles and the Tortoise"— attack the infinite divisibility of space. Take the dichotomy paradox. In order to complete any journey, you must first travel half the distance. But before you can do that, you must travel a quarter of the distance, and before that an eighth, and so on. In other words, you must complete an infinite number of sub-journeys in reverse order. So you can never get started.

A story has it that when Zeno told this paradox to Diogenes the Cynic, Diogenes "refuted" it by getting up and walking away. But Zeno's paradoxes are far from trivial. Bertrand Russell called them "immeasurably subtle and profound," and even today there is doubt among philosophers whether they have been completely resolved. Aristotle dismissed them as fallacies, but he was unable to disprove them; instead, he tried to block their conclusions by denying that there could be any actual infinity in nature. You could divide up space as finely as you pleased, Aristotle said, but you could never reduce it to an infinite number of parts.

Aristotle's abhorrence of the actual infinite came to pervade Greek thought, and a century later Euclid's *Elements* barred infinitesimal reasoning from geometry. This was disastrous for Greek science. The idea of the infinitely small had offered to bridge the conceptual gap between number and form, between the static and the dynamic. Consider the problem of finding the area of a circle. It is a straightforward matter to determine the area of a figure bounded by straight lines, such as a square or triangle. But how do you proceed when the boundary of the figure is curvilinear, as with a circle? The clever thing to do is to pretend the circle is a polygon made up of infinitely many straight-line segments, each of infinitesimal length. It was by approaching the problem in this way that Archimedes, late in the third century B.C.E., was able to establish the modern formula for circular area involving π. Owing to Euclid's strictures, however, Archimedes had to disavow his use of the infinite. He was forced to frame his demonstration as a reductio ad absurdum—a double reductio, no less—in which the circle was approximated by finite polygons with greater and greater numbers of sides. This cumbersome form of argument became known as the method of exhaustion, because it involved "exhausting" the area of a curved figure by fitting it with a finer and finer mesh of straight-edged figures.

For static geometry, the method of exhaustion worked well enough as an alternative to the forbidden infinitesimal. But it proved sterile in dealing with problems of dynamics, in which both space and time must be sliced to infinity. An object falling to earth, for example, is being continuously accelerated by the force of gravity. It has no fixed velocity for any finite interval of time, even one as brief as a thousandth of a second; every "instant" its speed is changing. Aristotle denied the meaningfulness of instantaneous speed, and Euclidean axiomatics could get no purchase on it. Only full-blooded infinitesimal reasoning could make sense of continuously accelerated motion. Yet that was just the sort of reasoning the Greeks fought shy of, because of the *horror infiniti* that was Zeno's legacy. Thus was Greek science debarred from attacking phenomena of matter in motion mathematically. Under Aristotle's influence, physics became a qualitative pursuit, and the

Pythagorean goal of understanding the world by number was abandoned. The Greeks might have amassed much particular knowledge of nature, but their love of rigor held them back from discovering a' single scientific law.

Though ostracized by Aristotle and Euclid, the infinitesimal did not entirely disappear from Western thought. Thanks to the enduring influence of Plato—who, unlike Aristotle, did not limit existence to what is found in the world of the senses—the infinitesimal continued to have a murky career as the object of transcendental speculation. Neoplatonists like Plotinus and early Christian theologians like Saint Augustine restored the infinite to respectability by identifying it with God. Medieval philosophers spent even more time engaged in disputation over the infinitely small than over the infinitely great.

With the revival of Platonism during the Renaissance, the infinitesimal began to creep back into mathematics, albeit in a somewhat mystical way. For Johannes Kepler, the infinitely small existed as a divinely given "bridge of continuity" between the curved and the straight. Untroubled by logical niceties—"Nature teaches geometry by instinct alone, even without ratiocination," he wrote—Kepler employed infinitesimals in 1612 to calculate the ideal proportions for that important thing, a wine cask. And his calculation was correct.

Kepler's friendliness toward the infinitesimal was shared by Galileo and Fermat. All three were edging away from the barren structure of Euclidean geometry toward a fertile, if freewheeling and unrigorous, science of motion, one that represented bodies as moving through infinitely divisible space and time. But there was a certain theological nettle to be grasped by these natural philosophers: How could the real infinite, which was supposed to be an attribute of God alone, be present in the finite world he created?

It was Blaise Pascal who was most agitated by this question. None of his contemporaries embraced the idea of the infinite more passionately than did Pascal. And no one has ever written with more conviction of the awe that the infinite vastness and minuteness of nature can evoke. Nature proposes the two infinities to us as mysteries, "not to understand, but to admire," Pascal wrote—and to use in our reasoning, he might have added. For Pascal was also a mathematician, and he freely introduced infinitely small quantities into his calculations of the areas of curvilinear forms. His trick was to omit them as negligible once the desired finite answer was obtained. This offended the logical sensibilities of contemporaries like Descartes, but Pascal replied to criticism by saying, in essence, that what reason cannot grasp, the heart makes clear.

Although Pascal's work prefigured the new science of nature, he (like Fermat and Galileo) never fully broke with the Euclidean tradition. But geometry alone was not up to the task of taming the infinitesimal, and if motion was to be understood quantitatively, the infinitesimal had to be tamed. This feat was finally achieved by Newton and Leibniz in the 1660s and '70s with their more or less simultaneous invention of the "calculus of infinitesimals"—which we now know simply as the calculus. The classical "geometrization" of nature gave way to its modern— and wildly successful—"mathematization," in the form of the calculus. The old philosophical perplexities about the infinitesimal were replaced by sheer wonder at its scientific fecundity.

And in Newton's hands, it could scarcely have been more fecund. Although his rival Leibniz worked out a more elegant formalism for the infinitesimal calculus—the same one in use today, in fact—it was Newton who used this new tool to bring a sense of harmony to the cosmos. Having framed his laws of motion and of gravity, he set out to deduce from them the exact nature of the orbit of a planet around the sun. This was a daunting task, given the continuous variation in a planet's velocity and distance from the sun. Instead of trying to arrive at the shape of the orbit all at once, Newton had the inspired idea of breaking it up into an infinite number of segments and then summing up the effects of the sun's gravitational force on the velocity of the planet in each infinitesimal segment.

Instantaneous velocity—a concept that had baffled Newton's predecessors—was defined as the ratio of two vanishingly small quantities: the infinitesimal distance traveled in an infinitesimal amount of time. Newton called such a ratio of vanishingly small quantities a "fluxion." Here's a simple illustration of how he used infinitesimals. Suppose a rock is dropped from the top of a building. In its descent to the ground, the rock is continuously accelerated by the earth's gravity. The distance it falls in t seconds is equal to $16t^2$ feet, so at the end of 1 second it will have fallen 16 feet ($= 16 \times 1^2$); at the end of 2 seconds it will have fallen 64 feet ($= 16 \times 2^2$); at the end of 3 seconds it will have fallen 144 feet ($= 16 \times 3^3$); and so on. Clearly, the rock's velocity is continuously increasing. Now suppose you want to know the rock's instantaneous velocity at some specific moment in its descent—at time t. According to Newton's reasoning, this instantaneous velocity will be a ratio of two infinitesimal quantities: an infinitesimal distance traveled just after t divided by an infinitesimal duration of time. So let's calculate this ratio, using ε to denote an infinitesimal bit of time. At t seconds, the rock will already have fallen $16t^2$ feet. An infinitesimal moment later, at $t + \varepsilon$, it will have fallen $16(t + \varepsilon)^2$ feet. So the distance it falls during the infinitesimal moment is the difference between these two distances: $16(t + \varepsilon)^2 - 16t^2$ feet. Expanding, you get $16t^2 + 32t\varepsilon + 16\varepsilon^2 - 16t^2$, which simplifies to $32\varepsilon + 16\varepsilon^2$ feet. To get the instantaneous velocity of the rock, you divide this infinitesimal distance by the infinitesimal time interval, which is just ε. The ratio of infinitesimals, then, is $(32t\varepsilon + \varepsilon^2)/\varepsilon$. Canceling the ε's in this ratio, you end up with $32t + 16\varepsilon$. But because the term 16ε in the final answer is infinitesimal (an infinitesimal multiplied by a finite number is still infinitesimal), it can be effectively set equal to zero—or so Newton reasoned. Therefore, the instantaneous velocity of the falling rock at time t is just $32t$ feet per second. Three seconds after it's dropped, the rock is falling at $32 \times 3 = 96$ feet per second.

From a far more sophisticated calculation in this vein, Newton was able to deduce that the planets should move in elliptical orbits with the sun at one focus—precisely the empirical law that Kepler had already formulated based on the voluminous sixteenth-century astronomical observations of Tycho Brahe. By dint of the infinitesimal calculus, Newton had managed to unify celestial and terrestrial motion.

Newton's demonstration of the law of ellipses was the single greatest achievement of the scientific revolution. The seeming implication—that nature obeys reason—made its discoverer the patron saint of the Enlightenment. Voltaire, after attending Newton's royal funeral in 1727, wrote, "Not long ago a distinguished company were discussing the trite and frivolous question: 'Who was the greatest man, Caesar, Alexander, Tamerlane, or Cromwell?' Someone answered that without doubt it was Isaac Newton. And rightly: for it is to him who masters our minds by the force of truth, not to those who enslave them by violence, that we owe our reverence." At a stroke Newton had transformed Aristotle's teleology-saturated cosmos into an orderly and rational machine, one that could serve the *philosophes* as a model for remaking human society. By elevating natural law to the status of objective fact, the Newtonian worldview inspired Thomas Jefferson's proposition that under the law of nature a broken contract authorized the Americans to rebel against George III.

Behind this triumph of human reason, however, lay an idea that still struck many as occult and untrustworthy. Newton himself was more than a little qualmish. In presenting his proof of the law of ellipses in the *Principia*, he purged it insofar as possible of the infinitesimal calculus; the resulting exposition, cast in a Euclidean mold, is impossible to follow. (Even the Nobel laureate Richard Feynman lost his way in the middle of Newton's argument when presenting it in a lecture to his students at Caltech.) In his later writings, Newton was careful never to consider infinitesimals in isolation but only in ratios, which were always finite. By the end of his life, he had renounced the idea of the infinitely small altogether.

Leibniz, too, had grave misgivings about the infinitesimals. On the one hand, they appeared to be required by his metaphysical principle *natura non facit saltum* (nature does not make leaps); without these amphibians traveling between existence and nonexistence, the transition from possibility to actuality seemed inconceivable. On the other hand, they resisted all attempts at rigorous definition. The best Leibniz could do was to multiply analogies, comparing, for instance, a grain of sand to the earth, and the earth to the stars. But when his pupil Johann Bernoulli cited the tiny creatures then being seen for the first time under the microscope (newly invented by Leeuwenhoek), Leibniz bridled, objecting that these animalcules were still of finite, not infinitesimal, size. Finally, he decided that infinitely small quantities were merely *fictiones bene fundatae* (well-founded fictions): they were useful to the art of discovery and did not lead to error, but they enjoyed no real existence.

For Bishop Berkeley, however, this was not good enough. In 1734, the philosopher published a devastating attack on the infinitesimal calculus titled *The Analyst; or, A Discourse Addressed to an Infidel Mathematician*. What motivated Berkeley was the threat to orthodox Christianity posed by the growing prestige of mechanistic science. (The "infidel mathematician" addressed is generally supposed to have been Newton's friend Edmund Halley.) As contrary to reason as the tenets of Christian theology might sometimes appear, Berkeley submitted, they were nowhere near so arcane and illogical as the linchpin of the new science, the infinitesimal. Defenders of the calculus were made to confront the following dilemma: either infinitesimals are exactly zero, in which case calculations involving division by them make no sense, or they are not zero, in which case the answers must be wrong. Perhaps, Berkeley derisively concluded, we are best off thinking of infinitesimals as "ghosts of departed quantities."

On the Continent, Voltaire, for one, was unbothered by scruples about the infinitely small, breezily describing the calculus as "the art of numbering and measuring exactly a thing whose existence cannot be conceived." As an instrument of inquiry, the infinitesimal calculus was simply too successful to be doubted. In the late eighteenth century, mathematicians like Lagrange and Laplace were using it to clear up even the difficult bits of celestial mechanics that had confounded Newton. The power of the calculus was matched by its versatility. It made possible the quantitative handling of all varieties of continuous change. The differential calculus showed how to represent the rate of change as a ratio of infinitesimals. The integral calculus showed how to sum up an infinite number of such changes to describe the overall evolution of the phenomenon in question. And the "fundamental theorem of calculus" linked these two operations in a rather beautiful way, by establishing that one was, logically speaking, the mirror image of the other.

During this golden age of discovery, scientists treated the infinitesimal as they would any other number, until it became convenient in their calculations to set it to zero (as Newton somewhat fishily did in cases like that of the falling rock, above). This cavalier attitude toward the infinitely small is captured by the advice of the French mathematician Jean Le Rond d'Alembert: "Allez en avant, et la foi vous viendra" (Go forward, and the faith will come to you).

Still, there remained those who felt it a scandal that the edifice of modern science was being erected on such metaphysically shaky foundations. Throughout the eighteenth century, there were many efforts to answer the charges against the infinitesimal put by critics like Berkeley and to find a logical set of rules for its use. None were successful;

some were simply fatuous. (Well into the next century, Karl Marx would try his hand at this problem, leaving nearly a thousand unpublished pages devoted to it.) One of the more philosophically appealing attempts was that of Bernard de Fontenelle, who tried to rationalize the infinitesimal by characterizing it as the reciprocal of the infinitely large. Though Fontenelle was ultimately defeated by formal difficulties, he was prescient in arguing that the reality of objects like the infinitesimal rested ultimately on their logical coherence, not on their existence in the natural world.

In the nineteenth century—by which time Hegel and his followers were seizing on confusions about the infinitesimal to support their contention that mathematics was self-contradictory—a way was finally found to get rid of this troublesome notion without sacrificing the wonderful calculus that was based on it. In 1821, the great French mathematician Augustin Cauchy took the first step by exploiting the mathematical notion of a "limit." The idea, which had been hazily present in the thought of Newton, was to define instantaneous velocity not as a ratio of infinitesimals but as the limit of a series of ordinary finite ratios; the members of this series, though never reaching the limit, come "as near as we please" to it. In 1858, the German mathematician Karl Weierstrass supplied a logically precise meaning to "as near as we please." Then, in 1872, Richard Dedekind, another German, showed how the smoothly continuous number line, previously thought to be held together by the glue of the infinitesimal, could be resolved into an infinity of rational and irrational numbers, no two of which actually touched.

All these developments were highly technical, and not a little painful to absorb. (They still are, as students of freshman calculus, made to struggle through mysterious "delta-epsilon" limit proofs, will tell you.) Taken together, they had three momentous consequences. First, they signaled the seemingly final ousting of the infinitely small from orthodox scientific thought. "There was no longer any need to suppose that there was such a thing," observed Bertrand Russell with evident relief. Second, they meant a return to Euclidean rigor for mathematics and its formal separation from physics after a heady era of discovery when the two were virtually indistinguishable. Third, they helped work a transformation in the prevailing philosophical picture of the world. If there is no such thing as the infinitesimal, then, as Russell observed, notions like "the next moment" and "state of change" become meaningless. Nature is rendered static and discontinuous, because there is no smooth transitional element to blend one event into the next. In a rather abstract sense, things no longer "hang together." The resulting sense of ontological discontinuity can be detected in the cultural lurch toward modernism—as witness Seurat's pointillism, Muybridge's stop-motion photography, the poetry of Rimbaud and Laforgue, the tone rows of Schoenberg, and the novels of Joyce.

A certain nostalgia for the infinitely small persisted among a few philosophical mavericks. Around the turn of the century, the French philosopher Henri Bergson argued that the new "cinematographic" conception of change falsified our pre-reflective experience, in which infinitesimal moments of time glided smoothly one into the next. In the United States, C. S. Peirce, one of the founders of pragmatism, similarly insisted on the primacy of our intuitive grasp of continuity. Peirce railed against the "antique prejudice against infinitely small quantities," arguing that the subjective *now* made sense only if interpreted as an infinitesimal. Meanwhile, in the mathematical world, the infinitesimal might have been expunged from "highbrow" mathematics, but it continued to be popular among "lowbrow" practitioners; physicists and engineers still found it an invaluable heuristic device in their workaday calculations—one that, for all its supposed muddledness, reliably led them to the right answer, as it had Newton.

After all, despite the strictures of Aristotle, Berkeley, and Russell, the infinitesimal had never been formally shown to be inconsistent. And with advances in logic during the early part of the twentieth century, a new understanding of consistency, and its relation to truth and existence, had begun to emerge. The prime mover was the Austrian-born logician Kurt Gödel. Today, Gödel is most famous for his "incompleteness theorem" of 1930, which says, roughly speaking, that no system of axioms is capable of generating all the truths of mathematics. In his doctoral thesis the year before, though, Gödel had proved a result of perhaps equal importance, which, somewhat confusingly, is known as the "completeness theorem." This theorem has a very interesting corollary. Take any set of statements that you please, couched in the language of logic. Then as long as those statements are mutually consistent—that is, as long as no contradiction can be deduced from them—the completeness theorem guarantees that there exists an interpretation in which those statements all come out as true. This interpretation is called a "model" for those statements. Consider, for example, the statements "all *a* are *b*" and "some *a* are *c*." If we interpret *a* as "humans," *b* as "mortal," and *c* as "redheaded," then the set of humans is a model for that pair of statements. Gödel showed how models could be constructed out of abstract mathematical ingredients. In doing so, he helped to inaugurate the field of logic called model theory, which studies the relationship between formal languages and their interpretations.

The most dramatic discovery that has been made by model theorists is that there is a fundamental indeterminacy in semantics—in the relationship between language and reality. A theory in a formal language, it turns out, is usually incapable of pinning down the unique reality it is supposed to describe. The theory will have "unintended interpretations" in which its meanings are twisted. To take a contrived example, consider the theory consisting of the single statement "all humans are mortal." Under its intended interpretation, the set of humans is a model for this theory. But if "humans" is taken to refer to cats, and "mortal" is taken to refer to the property of being curious, then the set of cats is also a model for this theory—an unintended one. A more interesting example is furnished by set theory. On the intended interpretation, the axioms of set theory describe an abstract universe of sets, and they entail the existence of higher infinities in that universe. But it turns out that those axioms can equally well be interpreted as being about the plain old counting numbers, among which no such higher infinities are to be found. So the axioms of set theory don't uniquely pin down the reality they are meant to describe. On one interpretation, they're about the universe of sets; on another interpretation, screwy yet equally valid, they're about the series 1, 2, 3, ... When we think we're saying true things about higher infinities, the noises we make could just as well be taken as expressing truths about ordinary numbers.

No one did more to exploit this fascinating indeterminacy than Abraham Robinson (1918–1974). For a logician, Robinson led a turbulent, yet urbane and even glamorous, life. Born in the Silesian mining village of Waldenburg (now Wałbrzych in Poland), he fled Nazi Germany with his family as a teenager. As a refugee in Palestine, Robinson joined the illegal Jewish militia called the Haganah while studying mathematics and philosophy at Hebrew University. A scholarship to the Sorbonne brought him to Paris shortly before it was taken by the Germans. Narrowly escaping, he managed to get to London during the Blitz and served as a sergeant for the Free French and then as a technical expert for the British air force. While pursuing pure mathematics and logic during the chaos of the war, Robinson also did brilliant work for the military in aerodynamics and "wing theory."

After the war, Robinson and his wife, a talented actress and fashion photographer from Vienna, could be found attending the haute couture collections together in Paris. Following teaching stints at the University of Toronto and Hebrew University, he was given Rudolf Carnap's old chair in philosophy and mathematics at UCLA at the beginning of the 1960s. Attracted by the lure of Hollywood, Robinson and his wife lived in a Corbusier-style villa in Mandeville Canyon, becoming friendly with the actor Oskar Werner. While doing work that made him one of the supreme mathematical logicians in the world, Robinson was also a convivial bon vivant as well as an early and vocal opponent of the Vietnam War. In the late 1960s, he moved to Yale, helping to transform it into a world center for logic before dying of pancreatic cancer in 1974, at the age of fifty-five.

Robinson's greatest feat of genius was his single-handed redemption of the infinitely small. He achieved this by thinking of the language of mathematics as a formal object, one that could be investigated and manipulated by logic. Here's the gist of how he proceeded. Start with the mathematical theory that describes how regular old arithmetic works—ordinary fractions, their addition and multiplication, and so forth. For brevity, let's call this theory of arithmetic T. We'll assume that T is a consistent theory, that there is no way to deduce within it a contradiction, like "$0 = 1$." (If ordinary arithmetic harbored contradictions, we'd be in real trouble: bridges would start falling down.)

Now let's add some stuff to the theory of arithmetic T. First, let's add a new symbol, which I'll call *INF* for "infinitesimal." Let's also add some new axioms that describe how *INF* is supposed to behave. We want *INF* to act like an infinitesimal: to be smaller than any finite number and yet larger than zero. We'll need a lot of new axioms to convey this—an infinite list of them, in fact. Here they are:

(New Axiom No. 1) *INF* is smaller than 1 but bigger than 0.

(New Axiom No. 2) *INF* is smaller than ½ but bigger than 0.

(New Axiom No. 3) *INF* is smaller than ⅓ but bigger than 0.

. . .

(New Axiom No. 1,000,000) *INF* is smaller than 1/1,000,000 but bigger than 0.

(New Axiom No. 1,000,001) *INF* is smaller than 1/1,000,001 but bigger than 0.

And so forth, endlessly.

Now let's use *T** to denote the enriched theory we get when we start with *T* and add all this new stuff. *T** seems to capture what we mean by the concept of the infinitesimal. It has a symbol, *INF*, designating a number that, according to the new axioms, is smaller than any finite number but bigger than zero. But how do we know that *T** is consistent? Remember, the great fear about the infinitesimal—a fear harbored by the Greeks, by Bishop Berkeley, and even by Newton—was that it might lead to paradox, inconsistency, contradiction. But Robinson succeeded in showing that this fear was unfounded. If *T*, the ordinary theory of arithmetic, is consistent, then *T**, the enriched theory embracing the infinitesimal, must also be consistent. That's what Robinson proved.

How did he do this? Well, suppose that *T** were inconsistent. Suppose, that is, that a contradiction could be proved from its axioms. This proof would consist of a finite number of lines—some of them axioms, some of them deductions from earlier lines—concluding with something absurd, like "$0 = 1$." Now, in this finite number of lines, only a finite number of the new axioms about *INF* could appear (at most one axiom per line). To make things vivid, let's say that the highest-numbered new axiom of *T** used in the proof happens to be

(New Axiom No. 147) *INF* is smaller than 1/147 but bigger than 0.

So we're supposing none of the new axioms of *T** beyond New Axiom No. 147 need to be invoked in the proof of the contradiction.

Now here comes the crucial move. Let's reinterpret *INF* as some plain-vanilla fraction that happens to be less than 1/147; specifically, let's say that *INF* is just a name for the fraction 1/148. Under this reinterpretation, none of the lines in the proof say anything about infinitesimals. They're all statements about ordinary fractions—perfectly true statements. So we now have on our hands a proof that is valid in the usual theory of arithmetic. But the last line of this proof is still "0 = 1." That means we have a proof that ordinary arithmetic is inconsistent! If the enriched theory T^* is inconsistent, then the ordinary theory T must be inconsistent too. Or, to turn the matter around, if the ordinary theory T is consistent, then the enriched theory T^* must also be consistent. So beefing up ordinary arithmetic by adding the stuff about infinitesimals doesn't raise the danger of inconsistency after all. The usual paradoxes associated with the infinitesimal are evaded because none of the new axioms, taken individually, says that *INF* is smaller than *all* the positive numbers. It takes the whole unending list of new axioms collectively to do that. But you can't get the whole list into a finite proof.

So we are safe, Robinson showed, in supposing that T^* is consistent. And there is more. With consistency in hand, we can invoke Gödel's completeness theorem, which says, in effect, *consistency is enough for reality*. A consistent theory is guaranteed to have a model: an abstract universe that the theory truly describes. And in the case of the enriched theory T^*, that model will be "nonstandard": it contains all sorts of exotic entities in addition to the ordinary finite numbers of arithmetic. Among the entities living in this nonstandard universe are infinitely small numbers. They surround each finite number in a tight little cloud that Robinson, in a nod to Leibniz, dubbed a "monad."

■

Robinson's epiphany about the infinitesimal came to him one day in 1961 as he walked into Fine Hall at Princeton, where he was visiting during a sabbatical. Five years later, he published *Non-standard Analysis*, in which he elaborated on the mathematical potential of his discovery. Robinson aptly chose the book's epigraph from Voltaire's story "Micromégas," about a giant extraterrestrial traveler who is amazed to encounter the relatively microscopic humans who inhabit the earth: "Je vois plus que jamais qu'il ne faut juger de rien sur sa grandeur apparente.

O Dieu! qui avez donné une intelligence à des substances qui paraissent si méprisables, l'infiniment petit vous coûte autant que l'infiniment grand." (I see more than ever that one should judge nothing by its apparent size. O God! who has given intelligence to such contemptible-seeming beings, the infinitely small costs you as little as the infinitely large.)

Curiously, adding infinitesimals to the universe of mathematics, as Robinson contrived to do, in no way alters the properties of ordinary finite numbers. Anything that can be proved about them using infinitesimal reasoning can, as a matter of pure logic, also be proved by ordinary methods. Yet this scarcely means that Robinson's approach is sterile. By restoring the intuitive methods that Newton and Leibniz pioneered, Robinson's "nonstandard analysis" has yielded proofs that are shorter, more insightful, and less ad hoc than their standard counterparts. Indeed, Robinson himself used it early on to solve a major open problem in the theory of linear spaces that had frustrated other mathematicians. Nonstandard analysis has since found many adherents in the international mathematical community, especially in France, and has been fruitfully applied to probability theory, physics, and economics, where it is well suited to model, say, the infinitesimal impact that a single trader has on prices.

Beyond his achievement as a mathematical logician, Robinson must be credited with bringing about one of the great reversals in the history of ideas. More than two millennia after the idea of the infinitely small had its dubious conception, and nearly a century after it had been got rid of seemingly for good, he managed to remove all taint of contradiction from it. Yet he did so in a way that left the ontological status of the infinitesimal completely open. There are those, of course, who believe that any mathematical object that does not involve inconsistency has a reality which transcends the world of our senses. Robinson himself subscribed to such a Platonistic philosophy early in his career, but he later abandoned it in favor of Leibniz's view that infinitesimals were merely "well-founded fictions."

Whatever reality the infinitesimal might have, it has no *less* reality than the ordinary numbers—positive, negative, rational and irrational, real and complex, and so on—do. When we talk about numbers, modern logic tells us, our language simply cannot distinguish between a nonstandard universe brimming with infinitesimals and a standard one that is devoid of them. Yet it remains a meaningful question whether the infinitesimal has physical reality—whether the infinitely small plays a role in the architecture of nature.

Might matter, space, and time be infinitely divisible? This bears on the timeless metaphysical question that was (reputedly) posed in fresh metaphorical form by Bertrand Russell: Is reality more like a pile of sand or a bucket of molasses? In this century, matter has been analyzed into atoms, which then turned out to consist of protons and neutrons, which in turn seem to be made up of smaller particles called quarks. Is that as far as it goes? Are these the material grains of sand? There is some evidence that quarks too have an internal structure, but probing that structure may require greater energies than physicists will ever be able to muster. As for space and time, according to current speculative theories they, too, could well have a discontinuous sand-like structure on the tiniest scale, with the minimum length being the Planck length of 10^{-33} centimeters and the minimum time being the Planck time of 10^{-43} seconds (exactly the time, it has been observed, that it takes a New York cabbie to honk after the light turns green). Again, though, proponents of infinite divisibility can always argue that with greater energies even smaller space-time scales could be detected, further worlds within worlds. They might also point to the "singularity" from which our universe was born in the big bang, an infinitely tiny point of energy. What better than the infinitesimal to serve as a principle of becoming, the ontological intermediary between being and nothingness?

Our most vivid sense of the infinitely small, however, may spring from our own finitude in the face of eternity, the thought of which can be at once humbling and ennobling. This idea, and its connection to the infinitely small, was expressed in a poignant way by Scott Carey, the protagonist of the 1950s film *The Incredible Shrinking Man*, as he seemed to be dwindling into nonexistence at the end of the movie, owing to the effect of some weird radiation. "I was continuing to shrink, to become—what?—the infinitesimal," he meditates, in a Pascalian vein, under the starry skies:

> So close, the infinitesimal and the infinite. But suddenly I knew they were really the two ends of the same concept. The unbelievably small and the unbelievably vast eventually meet, like

the closing of a gigantic circle. I looked up, as if somehow I could grasp the heavens. And in that moment I knew the answer to the riddle of the infinite. I had thought in terms of man's own limited dimensions. I had presumed upon nature. That existence begins and ends is man's conception, not nature's. And I felt my body dwindling, melting, becoming nothing. My fears melted away. And in their place came acceptance. All this vast majesty of creation—it had to mean something. And then I meant something too. Yes, smaller than the smallest, I meant something too. To God, there is no zero. I still exist.

And so, one feels, does the infinitesimal.

PART VI

Heroism, Tragedy, and the Computer Age

14

The Ada Perplex: Was Byron's Daughter the First Coder?

The programming language that the U.S. Department of Defense uses to control its military systems is named Ada, after Ada Byron, the daughter of Lord Byron. This act of nomenclature was not meant to be entirely fanciful. Augusta Ada Byron, who became by marriage the Countess of Lovelace, is widely supposed to have produced the first specimen of what would later be called computer programming. In her lifetime, she was deemed a mathematical prodigy, the Enchantress of Numbers. After her death in 1852—at the age of thirty-six, just like her father—popular biographers hymned her intellect and Byronic pedigree. With the coming of the computer age, Ada's posthumous renown expanded to new proportions. She has been hailed as a technological visionary; credited with the invention of binary arithmetic; made into the cult goddess of cyber feminism. It is an index of her prestige as a scientific virtuosa that Tom Stoppard, in his play *Arcadia*, presents a character based upon Ada groping her way toward the law of entropy and the theory of chaos, not to mention a proof of Fermat's last theorem.

The notion of Ada Lovelace as the inventor of computer programming appeals to the imagination because it contains two improbabilities and an irony. Improbability one: that computer programming, that seemingly masculine preserve, could have been invented by a woman. Improbability two: that the first program could have been written more than a century before a real computer came into existence. Irony: that this ur-programmer could have sprung from the loins of Lord Byron, who would have loathed anything having to do with a computer.

On reflection, there is something a little broad about that irony. Ada might have opted for algorithms over poems, but her very choice was a Byronic act of cosmic self-assertion, one founded on a highly romantic notion of her own genius. Like her father, she could be "mad and bad": hysteria prone and often opium addled, a compulsive gambler and lusty cocotte. "I am a d——d ODD animal," she observed of herself in a lucid moment. All of this is interesting. But so is another, quite independent question: Did Ada really play a vital role in the history of the computer?

The circumstances of Ada's birth, on December 10, 1815, were suitably melodramatic. On the eve of Ada's mother's labor, Byron spent the night in the room below hurling soda bottles at the ceiling. "Oh what an implement of torture have I acquired in you!" he is said to have exclaimed as he gazed at his newborn daughter. The marriage was only eleven months old, and in another month Lady Byron would leave her husband for good, taking Ada with her. Not only had she come to doubt his sanity, but she had also obtained evidence that he had been carrying on an incestuous relationship with his half sister, Augusta Leigh, and—even more damning in an era when sodomy was a capital crime—that he had engaged in homosexual acts. Newspapers rejoiced in the great separation drama, which became the archetype for the modern celebrity scandal. In April, Byron abandoned England for the Continent, choosing, in a final gesture of defiance, to be borne off to Dover in a replica of Napoleon's carriage. He never saw Ada again.

After the separation, Lady Byron devoted the rest of her life to two purposes: vindicating herself against Byron, and making sure that the notorious Byronic temperament did not emerge in her daughter. As a girl, Lady Byron had studied mathematics, for which her husband teasingly called her the Princess of Parallelograms. Now she decided that mathematics was just the thing to suppress any depraved tendencies Ada might have inherited from the paternal side. The little girl was put on a diet of sums and products and was allowed no contact with anything having to do with her father. She was eight when Byron died; the massive outpouring of grief that accompanied the return of his body to England barely touched her.

Ada did not flourish under her mother's regimen. At thirteen, she underwent an episode of hysterical blindness and paralysis. At sixteen, despite the constant surveillance of a coven of her mother's spinster friends ("the Furies," Ada called them), she managed to steal away for a spot of lovemaking with her tutor. To cool her passions, more mathematics was laid on, in the form of a six-volume primer on Euclid. Meanwhile, the pretty and flirtatious girl was besieged by suitors, who were attracted not only by her celebrity as a Byron but also by the enormous wealth she stood to inherit from her mother's richly landed family.

In 1833, at a party during her first London season, Ada was introduced to a forty-one-year-old widower named Charles Babbage. An accomplished mathematician, Babbage was also a prolific inventor of improving schemes and an all-purpose tinkerer. At the time, he was holding a series of soirees in his London home to show off something he called the Difference Engine: a mechanical calculating device, about the size of a traveling trunk, constructed of some two thousand glimmering brass and steel components—shafts, disks, gears—and powered by a hand crank. Ada went round with her mother to have a look at this "thinking machine" (as people were calling it) and was captivated. She asked Babbage for copies of plans and diagrams, which he gladly supplied.

Babbage had a practical motive for building the Difference Engine. With the coming of the Industrial Revolution, engineers and navigators needed accurate numerical tables; the ones they had contained thousands of errors, which could potentially result in shipwrecks and engineering disasters. In France, the director of the École Nationale des Ponts et Chaussées, Baron Gaspard Riche de Prony, had come up with an ingenious way of recalculating these tables when that nation went decimal, in 1799. De Prony found his inspiration in Adam Smith's *Wealth of Nations*—specifically, in Smith's account of the division of labor in a pin factory. Hiring a hundred or so Parisian hairdressers who had been thrown out of work when their clients lost their pompadoured heads to guillotines during the Reign of Terror, de Prony set up a kind of arithmetic assembly line that would, as he put it, "manufacture logarithms as one manufactures pins." The individual hairdressers had no special mathematical abilities: all they could do was add, subtract, and cut hair. The intelligence was in their organization. Babbage knew about de Prony's scheme from visits to Paris, and he came to the realization that these unskilled coiffeurs could be replaced by cogs. In other words, a machine could do the same calculation, automating mental effort just as the steam engine had automated physical effort.

Babbage's was not the first mechanical calculator. An adding machine had been invented in 1642 by Pascal (it was commercialized as the Pascaline), and in 1673 Leibniz contrived a machine capable of doing all four arithmetic operations, although the thing never worked well. No previous calculator, however, had been as intricately engineered as the Difference Engine, at least on the drawing board. The working model that Ada Byron saw, which had been constructed with the help of seventeen thousand pounds in grants from the British government (a sum sufficient to build two battleships), realized only a small part of Babbage's overall design. But after investing ten years of toil in the Difference Engine, Babbage chose to abandon it and start designing a far more ambitious machine. He called it the Analytical Engine.

The Analytical Engine was in many ways the prototype of the modern computer. Unlike the Difference Engine, whose material structure limited it to certain kinds of calculation, the Analytical Engine was to be truly programmable: depending on the instructions it was given, the same physical mechanism could carry out any mathematical function at all. (In modern terms, its "software" was to be independent of its "hardware.") Moreover, it would be able to alter its path during a computational run according to the outcome of intermediate calculations—in effect, performing an act of judgment using the logic of if-then. (Today, this is known as conditional branching.) Finally, the architecture of the Analytical Engine was quite like that of a modern computer: it had a "store" (memory), a "mill" (processor), and an input device for entering programs and an output device for printing results. The input device would read the programming instructions off punched cards, just as modern computers did until the late 1970s. Babbage borrowed the punched-card idea from French weaving technology; in 1804, Joseph-Marie Jacquard had invented a fully automated loom that would weave different patterns automatically depending on the sequence of punched cards fed into it.

Babbage worked out his design for the Analytical Engine in the years 1836–1840, all the while trying, in vain, to get government support for its realization. During the same period, Ada had a nervous breakdown; got married to a wellborn but rather stolid landowner who was later made the Earl of Lovelace; had three children and another nervous breakdown; got swept up in the craze for mesmerism and phrenology; became an avid harpist; and pursued her study of mathematics. Having finally been allowed to see a portrait of her notorious father and to read his poetry, and having been told by her mother of his probable incest and other forbidden experiments, Ada formed the idea of cleansing the Byronic legacy through science. "I have an ambition to make a compensation to mankind for his misused genius," she declared. "If he has transmitted to me any portion of that genius, I would use it to bring out great truths & principles. I think he has bequeathed this task to me!" She became more and more preoccupied with the genius she must certainly possess.

But how to express it? Despite her long immersion in mathematics, Ada's correspondence shows that she was still unable to do the most elementary trigonometry. (Her good friend Mary Somerville, by contrast, had already made original contributions to mathematics.) She realized that she would require still further instruction if she was to carry out her redemptive mission. So she began a search for "the desired Mathematical Man, the *Great Unknown*," finally obtaining the services of Augustus De Morgan, the first professor of mathematics at University College in London. After two years of his epistolary tuition (he was trying to teach her beginning calculus), Ada seems to have made little headway. In a letter to De Morgan dated November 27, 1842, she confesses to having struggled for eleven straight days over a problem that involved nothing more than substituting a simple mathematical expression into an equation. For her, the algebraic expressions in these student exercises were as trickily elusive as "sprites and fairies."

It was at this time, just as she was turning twenty-seven, that Ada at last hit upon the great undertaking that would focus her diffuse ambitions and, as it turned out, secure her posthumous fame. Two years earlier, Babbage had made the first and only public presentation of his blueprint for the Analytical Engine. The occasion was a scientific conference in Turin, to which Babbage had been invited as a foreign guest

of honor. Among the participants at this conference was Captain Lu-
igi Menabrea, a young military engineer who went on to become prime
minister of a newly unified Italy. Menabrea took notes on Babbage's
presentation, and in 1842 he published a paper, in French, with the title "Sketch of the Analytical Engine." A scientific friend of Ada's who
happened to see the paper suggested to her that she translate it for publication in a British scientific journal. Ada eagerly took on the project.
When she informed Babbage of what she was doing, he encouraged
her to add to the translation some notes of her own.

   This was an audacious proposal, for two reasons. First, women almost never published scientific papers in those days. Second, Ada was
not a professional scientist: she was a countess and a celebrity, the
daughter of the most famous literary figure of the day. For Babbage, a
relentless but not always effective self-promoter, the prospect of Ada
serving as his "interpretess" for the still-unbuilt-and-unfunded Analytical Engine must have been pulse-quickening, given the high circles she
moved in. Perhaps, he suggested, they should send a copy of her notes
to Prince Albert.

   Babbage provided Ada with ample assistance, not only explaining
the workings of his brainchild to her, but also supplying his own diagrams and formulas. The poetical conceits were Ada's alone. Stressing
that the Analytical Engine was to be programmable by punched cards,
like the automatic French looms, she wrote that it *weaves algebraical
patterns* just as the Jacquard-loom weaves flowers and leaves." Ada
gave several examples of the kinds of programs the machine would be
able to execute. In every case but one, these examples had been worked
out years earlier by Babbage, although, he later insisted, "the selection
was entirely her own."

   The one novel example in Ada's notes had to do with computing
what are known as the Bernoulli numbers. These numbers, first written
about in the previous century by the Swiss mathematician Jakob Bernoulli, crop up in the old-style calculation of navigational tables. Once
Babbage supplied Ada with the equation for them, she plunged into
the work of showing how it could be broken down into simpler formulas that could then be encoded as instructions for Babbage's machine.
But, with her shaky command of elementary algebra, her labor did
not go well. "I am in much dismay at having got into so amazing a

quagmire and botheration with these *Numbers*," she wrote to Babbage, who finally did the algebra himself "to save Lady Lovelace the trouble" (as he put it in his autobiography, written after her death). Ada then arranged Babbage's formulas in a "Table & diagram," similar in form to ones that he himself had created, which showed how each formula would be entered into the machine. The result, painstakingly inked in by her husband, Lord Lovelace, was published in August 1843. It is this proto-program (which, fittingly, contained a couple of proto-bugs) that has subsequently been used to justify the claim that Ada was the begetter of computer programming.

By the time it was finished, Ada's annotation had grown to some forty pages, more than twice the length of the translation it was supposed to be a gloss on. Although she described the overall architecture of the Analytical Engine, she showed scant concern with its mechanical details. She did, however, propose a number of philosophical opinions. She submitted that Babbage's machine transcended the merely numerical because it could operate on all kinds of symbols, not just numbers. It might, she wrote, be capable of composing "elaborate and scientific pieces of music of any degree of complexity or extent." She remarked that its operation would establish a link between matter and mental processes. Yet she insisted that unlike the human mind it could not be truly clever: "The Analytical Engine has no pretensions whatever to *originate* anything. It can do whatever we know how to order it to perform." Over a century later, Alan Turing dubbed this seeming truism "Lady Lovelace's objection" in a pioneering lecture he gave on artificial intelligence. What the objection overlooks is the possibility that a machine like Babbage's could alter its own instructions—in effect, learn from experience—to the point where it could do something both intelligent and unpredictable: like, say, checkmating a world chess champion.

It is doubtful whether Ada herself "originated" any of the ideas contained in her notes, except perhaps some of the more exuberantly speculative ones. On all technical and scientific points, regardless of how trifling, her letters show that she deferred to Babbage. Babbage, for his part, had good reason to connive in the fiction that the work was primarily Ada's: it not only made her notes a more effective piece of propaganda for his Analytical Engine but also enabled him to escape responsibility—on the pretense of not having been consulted—for some of her more hyperbolic claims. As for the other part of Ada's project—the translation of the Menabrea paper—that was marred by a highly embarrassing error, one that belies her reputation for mathematical competence. Owing to a typo in the original French edition, a phrase that should have been printed "le cas $n = \infty$" actually came out "le cos. $n = \infty$." Ada rendered this printer's lapse literally as "when the cos of $n = \infty$," which was obvious nonsense, because the cosine function always lies between +1 and −1.

Ada was well pleased with her work, which she began referring to as "this first child of mine." Upon its publication, her confidence in what she believed to be her extraordinary intellectual powers underwent further inflation. "I am in good spirits," she wrote to Babbage, "for I hope another year will make me *really* something of an *Analyst*. The more I study, the more irresistible do I feel my genius for it to be. I do *not* believe that my father was (or ever could have been) such a *Poet* as *I shall* be an *Analyst*, (& Metaphysician); for with me the two go together indissolubly." In keeping with her obsession with fairies and sprites—those mainstays of the mid-Victorian imagination—she urged him to surrender himself to her "Fairy-Guidance." Babbage declined to do so. The publication of Ada's notes had not been the coup he expected. The scientific establishment did not fall to its knees. The awaited parliamentary funding for constructing the Analytical Engine failed to materialize. Although Babbage spent the remaining twenty-eight years of his life promoting and refining his proto-computer, it never got off the drawing board.

It was Ada's own reputation that benefited most from the publication of her notes. Copies distributed to her society friends, including an actor, a playwright, and an art historian, elicited expressions of (bewildered) admiration. The coupling of this apparent scientific accomplishment—such a rarity for a woman—with her existing Byronic stardom made her the object of intense public interest. In 1844, London society buzzed with rumors that she was the anonymous author of a daring new book called *Vestiges of the Natural History of Creation*, which prefigured Darwin in depicting humans as the evolutionary product of a universe governed by natural laws. Ada had not written the book. In fact, other than an abortive review of a paper on animal magnetism, she was to write nothing else of any substance for the remainder of her life. Her lack of productivity stands in stark contrast with the scope of her ambitions. Among the achievements she now saw herself on the threshold of, the most fantastic was that of developing a "Calculus of the Nervous System": nothing less than a mathematical model of how the brain gives rise to thought.

To those who were intimate with her, such claims seemed mania-driven delusions of grandeur. "I hope the Self-esteem will not conduct its Owner to a Madhouse—it rather looks like it," Lady Byron remarked on her daughter's egomaniacal ravings. Like her father, Ada was subject to violent swings of mood that played havoc with her ability to focus on any single interest for very long. In addition to her recurrent nervous crises, she was plagued with a variety of debilitating physical problems, including gastric upsets, heart palpitations, asthma attacks, and some kind of kidney disorder. The all-purpose treatment for these complaints was opium, in the form of either laudanum (a solution of opium in wine spirits) or the newly discovered morphine. Ada depended on the "Opium system" for what tranquillity she could obtain. On one occasion, apparently under its influence, she had a vision of herself as the sun at the center of her own planetary system.

Ada's "Calculus of the Nervous System" never came to anything. But her preoccupation with electricity as the source of nervous energy did lead her to Andrew Crosse, an eccentric amateur scientist who was reputed to be Mary Shelley's model for Dr. Frankenstein, and thence to an adulterous affair with his son John. Indulging her Byronic appetite for forbidden experience, she also took to gambling on horses. There has been much speculation that Ada, as the ringleader of a group of punters, devised some kind of mathematical system for placing bets. If so, it was another illustration of her mathematical incompetence, for her turf losses became astronomical. She pawned the family diamonds to meet her gambling debts; Lady Byron quietly had them retrieved; Ada pawned them again. Her compulsive betting continued even after she was diagnosed with cancer of the womb. When massive hemorrhaging of blood confined her to bed, she was visited by Charles Dickens, who read to her from her favorite book, *Dombey and Son*. She was thirty-six when, after a prolonged period of agonizing throes, she died.

The other day, as I was looking at some of the many websites lionizing Ada, I came across the statement that "visits to Ada's grave now outnumber pilgrimages to the grave of Byron." This would be difficult to verify, because Ada, at her own deathbed request, was buried directly next to her father in the Byron vault in Nottingham. The same website referred to Ada as "the first hacker," which seems a desperate attempt to infuse a bit of glamour into what is today a very unglamorous avocation. Books about Ada abound in misinterpretations. Her first biographer, Doris Langley Moore, was a Byron specialist with no particular competence in mathematics. She describes Ada, in her "curious letters" to Augustus De Morgan, "enquiring, speculating, arguing, filling pages with equations, problems, solutions, algebraic formulae, like a magician's cabalistic symbols"—when in fact Ada was merely at the receiving end of a beginner's course in calculus. (A salutary exception is Ada's second biographer, Dorothy Stein, who deftly contrasted the grandiosity of Ada's aspirations with the modesty of her gifts and the slimness of her output.) The English academic Sadie Plant, in her cyber-feminist manifesto *Zeros+Ones* (1997), casually refers to "Ada's Analytical Engine," thereby dispossessing Babbage, the real creator of the first computer. Of course, Plant tells us, "Babbage's forward thinking was not a patch on Ada's own anticipative powers."

A later biography, Benjamin Woolley's *Bride of Science* (1999), is more sober in its assessment of Ada. Its author concedes that "Ada was not a great mathematician" and that she "probably did not have the mathematical knowledge to write the Menabrea notes without Babbage's help." But—perhaps because Woolley himself was a writer and broadcaster rather than a scientist—he does not dwell on Ada's technical prowess or lack thereof. Instead, he plays up her Byronic afflatus, saluting her as a practitioner of "poetical science." Through her lyrical similes—most famously, "the Analytical Engine *weaves algebraical patterns* just as the Jacquard-loom weaves flowers and leaves"—Ada "managed to rise above the technical minutiae of Babbage's extraordinary invention to reveal its true grandeur."

If Ada Lovelace did not invent computer programming, is it at least fair to say that Charles Babbage invented the computer? Woolley's biography of Ada makes much of the fact that Babbage never got his Analytical Engine built. The explanation, he argues, was not Bab-

bage's eccentric perfectionism or the limitations of contemporary engineering; it was that the Victorian world was simply not ready for
the computer. "All the areas of life, government and industry that the
computer has since revolutionized—telecommunications, administration, automation—barely existed at the time Ada translated and annotated the Menabrea notes," he writes. "When the electronic computer
emerged a hundred years later, its inventors knew very little about
Babbage and Ada."

Woolley is only half-right. It is true that there was no great practical role for computers in nineteenth-century life. A really compelling
need for them did not arise until World War II, when they proved decisive in breaking enemy codes. At that time, it was Alan Turing who
supplied the critical ideas. But Turing did know of Babbage's work;
in fact, his conception of a universal computing machine was very
close to Babbage's. The first digital computers of the early 1940s—
the Colossus at Bletchley Park in England, where thanks to Turing's
genius the Nazi Enigma cipher was cracked; the ENIAC (Electronic
Numerical Integrator and Computer) at the University of Pennsylvania; the Harvard Mark I, built by IBM—were all essentially Babbage
machines.

A question remains: Who was the first programmer? With Ada
Lovelace dismissed from contention, one might think that Babbage
merited this distinction, too, because he did write a number of programs
for his unrealized computer. But computers do not exhaust the universe
of programmable things. If "programming" means devising a set of
coded instructions that will get an automated contraption to do your
bidding, then the first great programmer was Joseph-Marie Jacquard—
the Frenchman who, at the beginning of the nineteenth century,
pioneered the use of punched cards to get automatic looms to weave
complicated patterns in brocade. Babbage himself acknowledged Jacquard's precedence: when he presented the concept for his Analytical
Engine at the Turin conference, he brought with him a silk portrait of
Jacquard that had been produced by an automatic loom programmed
by no fewer than twenty-four thousand cards. Even by today's standards, that's a lot of code.

Is it surprising that the earliest programs should be concerned not
with number crunching or information processing but with the weaving

of beautiful brocade? Or that the first functioning computer should consist not of mechanical components or vacuum tubes but of unemployed pompadour dressers? Such are the froufrou antecedents of the computer era—an era that can claim as its original publicist a nervy young woman, a poet's daughter, who saw herself as a fairy.

15

Alan Turing in Life, Logic, and Death

On June 8, 1954, Alan Turing, a forty-one-year-old research scientist at the University of Manchester, was found dead by his housekeeper. Before getting into bed the night before, he had taken a few bites out of an apple that was, apparently, laced with cyanide. At an inquest a few days later, his death was ruled a suicide. Turing was, by necessity rather than by inclination, a man of secrets. One of his secrets had been exposed two years before his death, when he was convicted of "gross indecency" for having a homosexual affair. Another, however, had not yet come to light. It was Turing who was chiefly responsible for breaking the German Enigma code during World War II, an achievement that helped save Britain from defeat in the dark days of 1941. Had this been publicly known, he would have been acclaimed a national hero. But the existence of the British code-breaking effort remained closely guarded even after the end of the war; the relevant documents weren't declassified until the 1970s. And it wasn't until the 1980s that Turing got the credit he deserved for a second, and equally formidable, achievement: creating the blueprint for the modern computer.

It is natural to view Turing as a gay martyr, hounded to death for his sexuality despite his great service to humanity. But it is also tempting to speculate about whether he really was a suicide. The flight to Moscow, in 1951, of Guy Burgess and Donald Maclean, British diplomats and rumored lovers who had been covertly working for the Soviets, prompted one London newspaper to editorialize that Britain should adopt the American policy of "weeding out both sexual and political perverts." Turing's role in wartime code breaking had left him with an intimate knowledge of British intelligence. After his conviction for homosexuality, he might have seemed out of control. He began traveling abroad in search of sex, visiting countries bordering on the Eastern bloc. The coroner at his inquest knew none of this. No one tested the apple found by his bedside for cyanide.

Could Turing have been the target of a clandestine assassination? The possibility has been raised more than once since his death, and it is hinted at by the title, borrowed from the Hitchcock thriller, of a short 2006 biography of Turing by David Leavitt, *The Man Who Knew Too Much*. Leavitt, the author of several novels and short-story collections with gay protagonists, rings the gay-martyr theme by invoking another film classic, *The Man in the White Suit*. In that 1951 comedy, which Leavitt reads as a gay allegory, a scientist is chased by a mob that feels threatened by a miraculous invention of his. Leavitt also mentions a third film, one that evidently made an impression on Turing: the 1937 Disney animation *Snow White and the Seven Dwarfs*. Those who knew Turing said that he was particularly fond of chanting the witch's couplet, "Dip the apple in the brew, / Let the sleeping death seep through."

Alan Mathison Turing was conceived in India, where his father worked in the Indian civil service, and born in 1912 during a visit by his parents to London. Instead of taking their child back to the East, they sent him to live with a retired army couple in a seaside English town. Alan was a good-looking boy, dreamy, rather clumsy, hopelessly untidy, and not very popular with his classmates. The loneliness of his childhood was finally dispelled when, in his early teens, he met another boy who shared his passion for science. They became inseparable friends, exploring esoterica like Einstein's relativity theory together. When, a year later, the boy died of tuberculosis, Turing seems to have been left with an ideal of romantic love that he spent the rest of his life trying to duplicate.

In 1931, Turing entered Cambridge. His college, King's, had (as Leavitt notes) "a very 'gay' reputation" and was known for its links to the Bloomsbury group. Turing's unworldliness kept him apart from the aesthetic set; he preferred the more Spartan pleasures of rowing and long-distance running. But Cambridge also had a rich scientific culture, and Turing's talents flourished in it. With the backing of John Maynard Keynes, he was elected a fellow of King's College in 1935, at the age of twenty-two. When the news reached his old school, the boys celebrated with a clerihew: "Turing / Must have been alluring / To get made a don / So early on." With a stipend, no duties, and high table dining privileges, he was free to follow his intellectual fancy.

That spring, attending lectures on the foundations of mathematics, Turing was introduced to a deep and unresolved matter known as the decision problem. A few months later, during one of his habitual runs, he lay down in a meadow and conceived a sort of abstract machine that settled it in an unexpected way.

The decision problem asks, in essence, whether reasoning can be reduced to computation. That was the dream of the seventeenth-century philosopher Gottfried von Leibniz, who imagined a calculus of reason that would permit disagreements to be resolved by taking pen in hand and saying, "Calculemus"—"Let us calculate." Suppose, that is, you have a set of premises and a putative conclusion. Is there some automatic procedure for deciding whether the former entails the latter—that is, whether the conclusion logically follows from the premises? Can you determine, in principle, whether a conjecture can be proved true or false? The decision problem calls for a mechanical set of rules for deciding whether such an inference is valid, one that is guaranteed to yield a yes-or-no answer in a finite amount of time. Such a method would be particularly useful to mathematicians, because it would allow them to resolve many of the conundrums in their field—like Fermat's last theorem, or Goldbach's conjecture—by brute force. That is why David Hilbert, who in 1928 challenged the mathematical community to solve the decision problem, called it "the principal problem of mathematical logic."

Turing began by thinking about what happens when a human carries out a computation by means of a pencil, a scratch pad, and a set of mindless instructions. By ruthlessly paring away inessential details, he arrived at an idealized machine that, he was convinced, captured the essence of the process. The machine was somewhat homely in conception: it consists of an unending tape divided into squares (rather like an infinite strip of toilet paper). Over this tape a little scanner travels back and forth, one square at a time, writing and erasing 0s and 1s, one per square. The scanner's action at any moment depends on the symbol in the square it is over and the state it is in—its "state of mind," so to speak. There are only a finite number of states, and the way they link up what the scanner sees to what it does constitutes the machine's program. (A typical line in a program would be something like "When the machine is in state A scanning 0, it will replace 0 by 1, move one square to the left, and then go into state B.")

Turing was able to do some amazing things with his abstract devices, which soon became known as Turing machines. Despite their simple design, he showed, they could be made to perform all sorts of complicated mathematics. Each machine's functioning, moreover, could be encapsulated in a single number (typically, a very long one), so that one machine could be made to operate on another by putting the number of the second machine on the tape of the first as a sequence of 0s and 1s. If a machine were fed its own number, then it could operate on itself. Turing was thereby able to exploit something akin to the paradoxes of self-reference ("I am lying") and show that certain sorts of Turing machines could not exist. For instance, there could be no Turing machine that, when fed with the program number of another machine, would decide whether that machine would eventually come to a halt in its computation or would grind on forever. (If there were such a machine, it could be tweaked into a Hamlet-like variant that would decide, in effect, "I will come to a halt if and only if I never come to a halt.") But the halting problem, it turned out, was merely the decision problem in disguise. Turing was able to prove that no computing machine of the kind he envisaged could solve the decision problem. Reasoning could not be reduced to computation after all.

But the death of Leibniz's dream turned out to be the birth of the computer age. The boldest idea to emerge from Turing's analysis was that of a universal Turing machine: one that, when furnished with the number describing the mechanism of any particular Turing machine, would perfectly mimic its behavior. In effect, the "hardware" of a special-purpose computer could be translated into "software" and then entered like data into the universal machine, where it would be run as a program. What Turing had invented, as a by-product of his advance in logic, was the stored-program computer.

Turing was twenty-three when he dispatched the decision problem. Just as he was finishing his work, discouraging news reached Cambridge from across the Atlantic: a Princeton logician named Alonzo Church had beaten him to the punch. Unlike Turing, however, Church did not arrive at the idea of a universal computing machine; instead, he used a far more arcane construction known as the lambda calculus. Still, Turing decided that he might profit from studying with the more established logician. So he made his way to America, crossing the Atlantic in steerage and arriving in New York, where, he wrote to his mother, "I had to go through the ceremony of initiation to the U.S.A., consisting of being swindled by a taxi-driver."

At Princeton, Turing took the first steps toward building a working model of his imaginary computer, pondering how to realize its logical design in a network of relay-operated switches; he even managed to get into a machine shop in the physics department and construct some of the relays himself. In addition to his studies with Church, he had dealings with the formidable John von Neumann, who would later be credited with innovations in computer architecture that Turing himself had pioneered. On the social side, he found the straightforward manners of Americans congenial, with certain exceptions: "Whenever you thank them for anything, they say 'You're welcome.' I rather liked it at first, thinking I was welcome, but now I find it comes back like a ball thrown against a wall, and become positively apprehensive. Another habit they have is to make the sound described by authors as 'Aha.' They use it when they have no suitable reply to a remark."

In 1938, Turing was awarded a Ph.D. in mathematics by Princeton and, despite the urgings of his father, who worried about imminent war with Germany, decided to return to Britain. Back at Cambridge, he became a regular at Ludwig Wittgenstein's seminar on the foundations of mathematics. Turing and Wittgenstein were remarkably alike: solitary, ascetic, homosexual, drawn to fundamental questions. But they disagreed sharply on philosophical matters, like the relationship between logic and ordinary life. "No one has ever yet got into trouble from a contradiction in logic," Wittgenstein insisted. To which Turing's response was "The real harm will not come in unless there is an application, in which case a bridge may fall down." Before long, Turing would himself demonstrate that contradictions could indeed have life-or-death consequences.

On September 1, 1939, Nazi troops invaded Poland. Three days later, Turing reported to Bletchley Park, a Victorian Tudor-Gothic estate northwest of London where the British cipher service had secretly relocated. He and the other code breakers arrived at Bletchley under the guise of "Captain Ridley's Shooting Party" (which had some locals grumbling about able-bodied men not doing their bit in the war). The task they faced was daunting. Since the use of radio communications in World War I, effective cryptography—ensuring that private messages could be sent via a public medium—had been critical to the military. The Nazis were convinced that their encryption system—based on a machine that looked like a souped-up typewriter, called the Enigma—would play a vital role in their expected victory.

The Enigma, invented for commercial use in 1918 and soon adopted by the German military, had an alphabetic keyboard and, next to that, a set of twenty-six little lamps, one for each letter. When a letter on the keyboard was pressed, a different letter on the lampboard would light up. If you typed the letters "d-o-g," the letters "r-l-u" might light up on the lampboard. When "rlu" was sent out in Morse code by a radio operator, a recipient would pick it up and type it on the keyboard of his Enigma machine, and the letters "d-o-g" would light up on the lampboard—so long as the settings of the two machines were the same. And that is where things get interesting. Inside the Enigma were a number of rotating wheels that determined the match between entered and coded letters; each time a letter was typed, one of the wheels would turn, altering the wiring. (Thus, if you typed "g-g-g," the coded version might be "q-d-a.") The military version of the Enigma also had something called a plugboard, by which the connections between letters could be further scrambled. The settings of the wheels and the plugboard were changed each day at midnight. And further layers of complexity were added, increasing the number of possible cipher keys to something like 150 quintillion.

The most impenetrable communications were those of the German navy, which used the Enigma machine with special cunning and discipline. By early 1941, Germany's growing U-boat fleet was devastating British shipping, sinking around sixty ships a month. Unlike Germany, Britain was almost completely reliant on the sea-lanes for sustenance. Unless some counterstrategy could be found, the British Isles faced being starved into submission. When Turing arrived at Bletchley Park, no work was being done on the naval Enigma, which many considered unbreakable. Indeed, it has been said there were only two people who thought the Enigma could be broken: Frank Birch, the head of Bletchley's naval-intelligence division, because it had to be broken; and Alan Turing, because it was an interesting problem.

Taking on the naval Enigma, Turing soon detected a weakness. A coded naval message would frequently contain formulaic bits, like WETTER FUER DIE NACHT (weather for the night), that might be guessed at. Such a "crib," he realized, could be exploited to yield logical chains, each of which corresponded to billions of possible Enigma settings. When one of these chains led to a contradiction—an internal inconsistency in a cipher hypothesis—the billions of settings to which it corresponded could be ruled out. Now the problem was reduced to checking millions of logical chains—daunting, to be sure, but not impossible. Turing set about devising a machine that would automate the search for logical consistency, eliminating contradictory chains rapidly enough for the code breakers to deduce that day's Enigma settings before the intelligence became stale. The result was the size of several refrigerators, with dozens of rotating drums (which mimicked the Enigma wheels) and massive coils of colored wire suggesting a Fair Isle sweater. In operation, it sounded like thousands of knitting needles clattering away as its relay switches checked one logical chain after another. In a nod to an earlier, Polish code-breaking machine, which made an ominous ticking sound, the people at Bletchley called the thing a Bombe.

On a good day, a Bombe could yield that day's Enigma key in as little as an hour, and by 1941 eighteen Bombes were up and running. With the Nazi naval communications rendered transparent, the British could pinpoint the position of the U-boats, steering convoys safely around them and, taking the offensive, sending destroyers to sink them. Even as the Battle of the Atlantic began to shift, the German High Command refused to believe that the Enigma could have been broken, suspecting instead espionage and treachery.

As the Enigma evolved, Turing continued to devise new strategies to defeat it. Known at Bletchley as the Prof, Turing was famed for his harmless eccentricities, like keeping his tea mug chained to the radiator and wearing a gas mask as he rode his bicycle to work (it helped to alleviate his hay fever). He impressed his colleagues as a friendly, approachable genius, always willing to explain his ideas, and he became especially close to a woman he worked with, playing what he called "sleepy chess" with her after their night-shift code breaking. Having convinced himself that he was in love, he proposed marriage and was eagerly accepted, even after he divulged his "homosexual tendencies" to her. But he later decided it wouldn't work and broke off the engagement. It seems to have been the only time in his life that he contemplated a heterosexual relationship.

By 1942, Turing had mastered most of the theoretical problems posed by the Enigma. Now that the United States was ready to throw its vast resources into the code-breaking effort, he was dispatched as a liaison to Washington, where he helped the Americans get their own Bombe making and Enigma monitoring under way. Then he headed to New York, where he was to work on another top secret project, involving the encryption of speech, at Bell Laboratories, which were then situated near the Hudson River piers in Greenwich Village. While at Bell Labs, he became engrossed with a question that came to occupy his postwar work: Was it possible to build an artificial brain? On one occasion, Turing stunned the entire executive mess at Bell Labs into silence by announcing, in a typically clarion tone, "I'm not interested in developing a powerful brain. All I'm after is just a mediocre brain, something like the president of the American Telephone and Telegraph Company."

Turing's early work had raised a fascinating possibility: perhaps the human brain is something like a universal Turing machine. Of course, the brain looks more like cold porridge than like a machine. But Turing suspected that what made the brain capable of thought was its logical structure, not its physical embodiment. Building a universal Turing machine might thus be the way to erase the line between the mechanical and the intelligent.

In 1945, Turing wrote up a plan for building a computer that contained everything from the abstract structure down to the circuit diagrams and a cost estimate of 11,200 pounds. At Britain's National Physical Laboratory, where he worked after the war, he had nothing like the resources of the Americans, and yet he rose to the challenge posed by his straitened circumstances. When it came to the computer's memory, for example, the most obvious storage device was one in which the data took the form of vibrations in liquid mercury. But Turing reckoned that gin would be just as effective and far cheaper. On one occasion, he noticed a drainpipe lying in a field and had a colleague help him drag it back to the laboratory for use in his computer hardware. Frustrated with the inept administration at the NPL, he finally accepted an offer to direct the development of a computer prototype at the University of Manchester. Arriving in that grim northern industrial city at the age of thirty-six, he found it "mucky" and noted that the Mancunian male wasn't much to look at.

Despite his immersion in engineering details, Turing's fascination with computing was essentially philosophical. "I am more interested in the possibility of producing models of the action of the brain than in the practical applications of computing," he wrote to a friend. Turing conjectured that, initially at least, computers might be suited to purely symbolic tasks, those presupposing no "contact with the outside world," like mathematics, cryptanalysis, and chess playing (for which he himself worked out the first programs on paper). But he imagined a day when a machine could simulate human mental abilities so well as to raise the question of whether it was actually capable of thought. In a paper published in the philosophy journal *Mind*, he proposed the now classic "Turing test": a computer could be said to be intelligent if it could fool an interrogator—perhaps in the course of a dialogue conducted via Teletype—into thinking it was a human being. Turing argued that the only way to know that other people are conscious is by comparing their behavior with one's own and that there is no reason to treat machines any differently.

To David Leavitt, the idea of a computer mimicking a human inevitably suggests that of a gay man "passing" as straight. Leavitt shows a rather overdeveloped ability to detect psychosexual significance. (When, in the *Mind* paper, Turing writes of certain human abilities that it is hard to imagine a machine developing, like the ability to "enjoy strawberries and cream," Leavitt sees a "code word for tastes that Turing prefers not to name.") But Leavitt does succeed, on the whole, in giving a poignant depiction of Turing the man. It is on the technical side that Leavitt falls short. His exposition, full of the sort of excess detail that mathematicians call "hair," is marred by confusions and errors.

In trying to describe how Turing resolved the decision problem, Leavitt gets wrong the central idea of a "computable number." Discussing the earlier logical work of Kurt Gödel, Leavitt says that it established that the axiomatic system of Bertrand Russell and Alfred North Whitehead's *Principia Mathematica* was "inconsistent," when Gödel proved no such thing, and a definition of something called the Skewes number is precisely backward. Although Leavitt seems to have made a valiant attempt to master this material in preparation for writing the book, his explanatory efforts will leave initiates irritable and beginners perplexed.

In fairness, the bar for Leavitt had been set pretty high. In 1983, a mathematician named Andrew Hodges published *Alan Turing: The Enigma*, which is one of the finest scientific biographies ever written and has remained an essential resource for all subsequent accounts of Turing's life. In 1987, Hugh Whitemore's superb play about Turing, *Breaking the Code*, opened on Broadway, with Derek Jacobi in the starring role. Both of these works not only captured the pathos of Turing's life; they also gave a lucid account of his technical achievement. Whitemore's play miraculously compressed the decision problem and the Enigma decoding into a couple of brief speeches without any real distortion. (By contrast, the 2014 film *The Imitation Game*, starring Benedict Cumberbatch in the role of Turing, took intolerable liberties with the details of both Turing's life and his code-breaking achievement, depicting a man who was by all accounts forthright, witty, and generously collegial as a humorless and even timorous nerd.)

Turing lived for the remainder of his life in Manchester. He bought a small house in a suburb and bicycled the ten miles to the university each day, donning a slightly ludicrous yellow oilskin and hat when it rained. Although nominally the deputy director of the computing laboratory (which developed the world's first commercially available electronic computer), he also took on a fundamental mystery in biology: How is it that living things, which start out as a cluster of identical cells, eventually grow into the variety of different forms found in nature? Working out systems of equations to model this process of morphogenesis, he used the prototype computer to find solutions; seated at the console, using the machine's manual controls, Turing looked, in the words of one colleague, as if he were "playing the organ."

Shortly before Christmas 1951, Turing was walking along Oxford Street in Manchester when his eye was caught by a nineteen-year-old working-class youth named Arnold Murray. The encounter turned into an affair of sorts, with Murray coming to Turing's house on several occasions, having dinner with him, and then spending the night. A month later, Turing was invited by the BBC to take part in a radio debate on the question "Can automatic calculating machines be said to think?" (He had already received some rather breathless publicity on his ideas about artificial intelligence from the British papers.) On one of the days that the program aired, Turing came home to find that his house had been burglarized. The burglar, as he suspected, was an associate of Murray's who was confident that a homosexual would never go to the police.

But Turing did go to the police. After some initial dissembling about how he came by his information about the culprit's identity, Turing volunteered the details of his affair to the startled detectives. Turing was charged, under the same 1885 act that led to the prosecution of Oscar Wilde, with "gross indecency." This crime was punishable by up to two years' imprisonment, but the judge, taking into account Turing's intellectual distinction (though knowing nothing of his activities during the war), sentenced him to probation, on the condition that he "submit for treatment by a duly qualified medical practitioner."

The treatment of choice was hormonal. Earlier, American researchers had tried to convert gay men to heterosexuality by injecting them with male hormones, on the theory that they suffered from a masculinity deficit; surprisingly, this only seemed to intensify their homosexual drive. So the opposite approach was tried. By giving homosexuals large doses of female hormones, it was found, their libido could be destroyed in as little as a month. This chemical castration had the side effect of causing temporary breast enlargement, as Turing found to his humiliation, and his lean runner's body took on fat.

The news of Turing's conviction received no national attention. The reaction of his mother, to whom he had grown close over the years, was one of affectionate exasperation. His lab colleagues dismissed it all as "typical Turing." With his criminal record of "moral turpitude," he was barred from the United States. But, once his probation ended, in April 1953, and the effects of the hormone regimen wore off, he traveled to Europe for romantic liaisons. His position at Manchester was secure: the university created a special Readership in the Theory of Computing for him, which came with a pay raise. He was free to continue with his work on mathematical biology and artificial intelligence, and he enjoyed the growing talk among logicians of "Turing machines."

Why, then, more than two years after the trial, and more than a year after the hormone treatment ended, would he have committed suicide? Leavitt describes Turing's life after his arrest as "a slow, sad descent into grief and madness." That's overly dramatic. Turing did start seeing a Jungian analyst and developed a taste for Tolstoy, but neither is an infallible sign of madness. He also, a few months before his death, sent a friend a series of postcards containing eight "messages from the unseen world." Some were terse aphorisms: "Science is a differential Equation. Religion is a Boundary Condition." Others had a Blakean cast: "Hyperboloids of wondrous Light / Rolling for aye through Space and Time / Harbour those Waves which somehow might / Play out God's wondrous pantomime." Well, it does rhyme.

Turing's death occurred in a period of acute anxiety about spies and homosexuals and Soviet entrapment. That week, newspapers announced that the former head of Los Alamos, Robert Oppenheimer, had been judged a security risk. And, as Andrew Hodges wrote, "had the headline been 'ATOMIC SCIENTIST FOUND DEAD,' the questions would have been immediate and public." Still, there is no direct evidence that the death of the Man Who Knew Too Much was anything other than a suicide. Indeed, the only person who seems to have had doubts was Turing's mother, who insisted that her son must have accidentally ingested something from one of the chemical experiments he conducted at home. Turing was rather sloppy, and he was known to eat an apple every night before going to bed. On the other hand, he once wrote a letter to a friend mentioning a method of suicide that "involved an apple and electric wiring."

Was Turing's death a kind of martyrdom? Was it the perfect suicide—one that deceived the person whose feelings he cared most about, his mother—or, more improbably, the perfect murder? These questions have been repeatedly raised over the years, by Leavitt and others, yet they remain unresolved. Perhaps, Leavitt invites us to imagine, the message Turing wanted to convey is one that has so far been overlooked: "In the fairy tale the apple into which Snow White bites doesn't kill her; it puts her to sleep until the Prince wakes her up with a kiss." This note of macabre camp doesn't suit a man who eschewed all forms of egoistic fuss as he solved the most important logic problem of his time, saved countless lives by defeating a Nazi code, conceived the computer, and rethought how mind arises from matter.

16

Dr. Strangelove Makes
a Thinking Machine

The digital universe came into existence, physically speaking, late in 1950, in Princeton, New Jersey, at the end of Olden Lane. That was when and where the first genuine computer—a high-speed, stored-program, all-purpose digital-reckoning device—stirred into action. It had been wired together, largely out of military surplus components, in a one-story cement-block building that the Institute for Advanced Study had constructed for the purpose. The new machine was dubbed MANIAC, an acronym of "Mathematical and Numerical Integrator and Computer."

And what was MANIAC used for, once it was up and running? Its first job was to do the calculations necessary to engineer the proto-type of the hydrogen bomb. Those calculations were successful. On the morning of November 1, 1952, the bomb they made possible, nick-named Ivy Mike, was secretly detonated over a South Pacific island called Elugelab. The blast vaporized the entire island, along with eighty million tons of coral. One of the air force planes sent in to sample the mushroom cloud—reported to be "like the inside of a red-hot furnace"—spun out of control and crashed into the sea; the pilot's body was never found. A marine biologist on the scene recalled that a week after the H-bomb test he was still finding terns with their feathers blackened and scorched and fish whose "skin was missing from a side as if they had been dropped in a hot pan."

The computer, one might well conclude, was conceived in sin. Its birth helped ratchet up, by several orders of magnitude, the destructive force available to the superpowers during the cold war. And the man most responsible for the creation of that first computer, John von Neumann, was himself among the most ardent of the cold warriors, an advocate of a preemptive military attack on the Soviet Union, and one of the models for the film character Dr. Strangelove. "The digital universe and the hydrogen bomb were brought into existence at the same time," the historian of science George Dyson has observed. Von Neumann had seemingly made a deal with the devil: "The scientists would get the computers, and the military would get the bombs." And many scientists at the institute were by no means happy with this deal—including one who wrote STOP THE BOMB in the dust on von Neumann's car.

It was not just the military impetus behind the project that evoked opposition at the institute. Many felt that such a number-crunching behemoth, whatever its purpose, had no place in what was intended to be a sort of Platonic heaven for pure scholarship. The Institute for Advanced Study was founded in 1930 by the brothers Abraham and Simon Flexner, philanthropists and educational reformers. The money came from Louis Bamberger and his sister Caroline Bamberger Fuld, who sold their interest in the Bamberger's department store chain to Macy's in 1929, just weeks before the stock market crash. Of the eleven million dollars of the proceeds they took in cash, the Bambergers committed five million dollars (equivalent to sixty million dollars today) to establishing what Abraham Flexner envisaged as "a paradise for scholars who, like poets and musicians, have won the right to do as they please." The setting was to be Olden Farm in Princeton, the site of a skirmish during the Revolutionary War.

Although some thought was given to making the new institute a center for economics, the founders decided to start with mathematics, because of both its universal relevance and its minimal material requirements: "a few rooms, books, blackboard chalk, paper, and pencils," as one of the founders put it. The first appointee was Oswald Veblen (Thorstein Veblen's nephew) in 1932, followed by Albert Einstein—who, on his arrival in 1933, found Princeton to be "a quaint and ceremonious little village of puny demigods on stilts" (or so at least he told the queen of Belgium). That same year, the institute hired John von

Neumann, a Hungarian-born mathematician who had just turned twenty-nine.

Among twentieth-century geniuses, von Neumann ranks very close to Einstein. Yet the styles of the two men were quite different. Whereas Einstein's greatness lay in his ability to come up with a novel insight and elaborate it into a beautiful (and true) theory, von Neumann was more of a synthesizer. He would seize on the fuzzy notions of others and, by dint of his prodigious mental powers, leap five blocks ahead of the pack. "You would tell him something garbled, and he'd say, 'Oh, you mean the following,' and it would come back beautifully stated," said his onetime protégé the Harvard mathematician Raoul Bott.

Von Neumann might have missed Budapest's café culture in provincial Princeton, but he felt very much at home in his adopted country. Having been brought up a Hungarian Jew in the late Hapsburg Empire, he had experienced the short-lived Communist regime of Béla Kun after World War I, which made him, in his words, "violently anti-Communist." After returning to Europe in the late 1930s to court his second wife, Klári, he finally quit the Continent with an implacable enmity for the Nazis, growing suspicions of the Soviets, and (in the words of George Dyson) "a determination never again to let the free world fall into a position of military weakness that would force the compromises that had been made with Hitler." His passion for America's open frontiers extended to a taste for large, fast cars; he bought a new Cadillac every year (whether he had wrecked the last one or not) and loved speeding across the country on Route 66. He dressed like a banker, gave lavish cocktail parties, and slept only three or four hours a night. Along with his prodigious intellect went (according to Klári) an "almost primitive lack of ability to handle his emotions."

It was toward the end of World War II that von Neumann conceived the ambition of building a computer. He spent the latter part of the war working on the atomic bomb project at Los Alamos, where he had been recruited because of his expertise at the (horribly complicated) mathematics of shock waves. His calculations led to the development of the "implosion lens" responsible for the atomic bomb's explosive chain reaction. In doing them, he availed himself of some mechanical tabulating machines that had been requisitioned from

IBM. As he familiarized himself with the nitty-gritty of punch cards and plugboard wiring, the once pure mathematician became engrossed by the potential power of such machines. "There already existed fast, automatic special purpose machines, but they could only play one tune . . . like a music box," said Klári, who had come to Los Alamos to help with the calculations; "in contrast, the 'all purpose machine' is like a musical instrument."

As it happened, a project to build such an "all purpose machine" had already been secretly launched during the war. It was instigated by the army, which desperately needed a rapid means of calculating artillery-firing tables. (Such tables tell gunners how to aim their weapons so that the shells land in the desired place.) The result was a machine called ENIAC (for "Electronic Numerical Integrator and Computer"), built at the University of Pennsylvania. The co-inventors of ENIAC, John Presper Eckert and John Mauchly, cobbled together a monstrous contraption that, despite the unreliability of its tens of thousands of vacuum tubes, succeeded at least fitfully in doing the computations asked of it. ENIAC was an engineering marvel. But its control logic—as von Neumann soon saw when he was given clearance to examine it—was hopelessly unwieldy. "Programming" the machine involved technicians spending tedious days reconnecting cables and resetting switches by hand. It thus fell short of the modern computer, which stores its instructions in the form of coded numbers, or "software."

Von Neumann aspired to create a truly universal machine, one that (as Dyson aptly puts it) "broke the distinction between numbers that mean things and numbers that do things." A report sketching the architecture for such a machine—still known as the von Neumann architecture—was drawn up and circulated toward the end of the war. Although the report contained design ideas from the ENIAC inventors, von Neumann was listed as the sole author, which occasioned some grumbling among the uncredited. And the report had another curious omission. It failed to mention the man who, as von Neumann well knew, had originally worked out the possibility of a universal computer: Alan Turing.

An Englishman nearly a decade younger than von Neumann, Alan Turing came to Princeton in 1936 to earn a Ph.D. in mathematics. Earlier that year, at the age of twenty-three, he had resolved a deep problem in logic called the decision problem. The problem traces its origins to the seventeenth-century philosopher Leibniz, who dreamed of "a universal symbolistic in which all truths of reason would be reduced to a kind of calculus." Could reasoning be reduced to computation, as Leibniz imagined? More specifically, is there some automatic procedure that will decide whether a given conclusion logically follows from a given set of premises? That was the decision problem. And Turing answered it in the negative: he gave a mathematical demonstration that no such automatic procedure could exist. In doing so, he came up with an idealized machine that defined the limits of computability: what is now known as a Turing machine.

The genius of Turing's imaginary machine lay in its stunning simplicity. ("Let us praise the uncluttered mind," exulted one of Turing's colleagues.) It consisted of a scanner that moved back and forth over an infinite tape reading and writing 0s and 1s according to a certain set of instructions—0s and 1s being capable of expressing all letters and numerals. A Turing machine designed for some special purpose—like adding two numbers together—could itself be described by a single number that coded its action. The code number of one special-purpose Turing machine could even be fed as an input onto the tape of another Turing machine. This led Turing to the idea of a *universal* machine: one that, if fed the code number of any special-purpose Turing machine, would function as if it actually were that special-purpose machine. For instance, if a universal Turing machine were fed the code number of the Turing machine that performed addition, the universal machine would temporarily turn into an adding machine. That is exactly what happens when your laptop, which is a physical embodiment of Turing's universal machine, runs a word-processing program or when your smartphone runs an app. Thus did Turing create the template for today's stored-program computer.

When Turing subsequently arrived at Princeton as a graduate student, von Neumann made his acquaintance. "He knew all about Turing's work," said a co-director of the computer project. "The whole relation of the serial computer, tape and all that sort of thing, I think was very clear—that was Turing." Von Neumann and Turing were virtual opposites in character and appearance: the older man a portly, well-attired, and clubbable sybarite who relished wielding power and influence; the younger one a shy, slovenly, dreamy ascetic (and homosexual), fond of intellectual puzzles, mechanical tinkering, and long-distance running. Yet the two shared a knack for getting to the logical essence of things. After Turing completed his Ph.D. in 1938, von Neumann offered him a salaried job as his assistant at the institute, but with war seemingly imminent Turing decided to return to England instead.

"The history of digital computing," Dyson writes in his 2012 book, *Turing's Cathedral*, "can be divided into an Old Testament whose prophets, led by Leibniz, supplied the logic, and a New Testament whose prophets, led by von Neumann, built the machines. Alan Turing arrived in between." It was from Turing that von Neumann drew the insight that a computer is essentially a logic machine—an insight that enabled him to see how to overcome the limitations of ENIAC and realize the ideal of a universal computer. With the war over, von Neumann was free to build such a machine. And the leadership of the Institute for Advanced Study, fearful of losing von Neumann to Harvard or IBM, obliged him with the authorization and preliminary funding.

There was widespread horror among the institute's fellows at the prospect of such a machine taking shape in their midst. The pure mathematicians tended to frown on tools other than blackboard and chalk, and the humanists saw the project as mathematical imperialism at their expense. "Mathematicians in our wing? Over my dead body! and yours?" a paleographer cabled the institute's director. (It didn't help that fellows already had to share their crowded space with remnants of the old League of Nations that had been given refuge at the institute during the war.) The subsequent influx of engineers raised hackles among both the mathematicians and the humanists. "We were doing things with our hands and building dirty old equipment. That wasn't the institute," one engineer on the computer project recalled.

Von Neumann himself had little interest in the minutiae of the computer's physical implementation; "he would have made a lousy engineer," said one of his collaborators. But he recruited a resourceful team, led by the chief engineer, Julian Bigelow, and he proved a shrewd manager. "Von Neumann had one piece of advice for us," Bigelow recalled: "not to originate anything." By limiting the engineers to what was strictly needed to realize his logical architecture, von Neumann saw to it that MANIAC would be available in time to do the calculations critical for the hydrogen bomb.

The possibility of such a "superbomb"—one that would, in effect, bring a small sun into existence without the gravity that keeps the sun from flying apart—had been foreseen as early as 1942. If a hydrogen bomb could be made to work, it would be a thousand times as powerful as the bombs that destroyed Hiroshima and Nagasaki. Robert Oppenheimer, who had led the Los Alamos project that produced those bombs, initially opposed the development of a hydrogen bomb on the grounds that its "psychological effect" would be "adverse to our interest." Other physicists, like Enrico Fermi and Isidor Rabi, were more categorical in their opposition, calling the bomb "necessarily an evil thing considered in any light." But von Neumann, who feared that another world war was imminent, was enamored of the hydrogen bomb. "I think that there should never have been any hesitation," he wrote in 1950, after President Truman decided to proceed with its development.

Perhaps the fiercest advocate of the hydrogen bomb was the Hungarian-born physicist Edward Teller, who, backed by von Neumann and the military, came up with an initial design. But Teller's calculations were faulty; his prototype would have been a dud. This was first noticed by Stanislaw Ulam, a brilliant Polish-born mathematician (elder brother to the Sovietologist Adam Ulam). Having shown that the Teller scheme was a nonstarter, Ulam produced, in his typically absentminded fashion, a workable alternative. "I found him at home at noon staring intensely out of a window with a very strange expression on his face," Ulam's wife recalled. "I can never forget his faraway look as peering unseeing in the garden, he said in a thin voice— I can still hear it—'I found a way to make it work.'"

Now Oppenheimer—who had been named director of the Institute for Advanced Study after leaving Los Alamos—was won over. What became known as the Teller-Ulam design for the H-bomb was, Oppenheimer said, "technically so sweet" that "one had to at least make the thing." And so, despite strong opposition on humanitarian grounds among many at the institute (who suspected what was going on from the armed guards stationed by a safe near Oppenheimer's office), the newly operational computer was pressed into service. The thermonuclear calculations kept it busy for sixty straight days, around the clock, in the summer of 1951. MANIAC did its job perfectly. Late the next year, "Ivy Mike" exploded in the South Pacific, and Elugelab island was removed from the map.

Shortly afterward, von Neumann had a rendezvous with Ulam on a bench in Central Park, where he probably informed Ulam firsthand of the secret detonation. But then (judging from subsequent letters) their chat turned from the destruction of life to its creation, in the form of digitally engineered self-reproducing organisms. Five months later, the discovery of the structure of DNA was announced by Francis Crick and James Watson, and the digital basis of heredity became apparent. Soon MANIAC was being given over to problems in mathematical biology and the evolution of stars. Having delivered its thermonuclear calculations, it became an instrument for the acquisition of pure scientific knowledge, in keeping with the purpose of the institute where it was created.

But in 1954, President Eisenhower named von Neumann to the Atomic Energy Commission, and with his departure the institute's computer culture went into decline. Two years later, the fifty-two-year-old von Neumann lay dying of bone cancer in Walter Reed Army Hospital, disconcerting his family by converting to Catholicism near the end. (His daughter believed that von Neumann, an inventor of game theory, must have had Pascal's wager in mind.) "When von Neumann tragically died, the snobs took their revenge and got rid of the computing project root and branch," the physicist Freeman Dyson later commented, adding that "the demise of our computer group was a disaster not only for Princeton but for science as a whole." At exactly midnight on July 15, 1958, MANIAC was shut down for the last time. Its corpse now reposes in the Smithsonian Institution in Washington.

Was the computer conceived in sin? The deal von Neumann made with the devil proved less diabolical than expected. As George Dyson observes, "It was the computers that exploded, not the bombs." Yet it is interesting to consider how von Neumann's vision of the digital future has been superseded by Turing's. Instead of a few large machines handling the world's demand for high-speed computing, as von Neumann envisaged, a seeming infinity of much smaller devices, including the billions of microprocessors in cell phones, have coalesced into what Dyson calls "a collective, metazoan organism whose physical manifestation changes from one instant to the next." And the progenitor of this virtual computing organism is Turing's universal machine.

So the true dawn of the digital universe came not in the 1950s, when von Neumann's machine started running thermonuclear calculations. Rather, it was in 1936, when the young Turing, lying down in a meadow during one of his habitual long-distance runs, conceived of his abstract machine as a means of solving a problem in pure logic. Like von Neumann, Turing was to play an important behind-the-scenes role in World War II. Working as a code breaker for his nation at Bletchley Park, he deployed his computational ideas to crack the Nazi Enigma code, an achievement that helped save Britain from defeat in 1941 and reversed the tide of the war. But Turing's wartime heroism remained a state secret well beyond his suicide in 1954, two years after he had been convicted of "gross indecency" for a consensual homosexual affair and sentenced to chemical castration.

In 2009, the British prime minister, Gordon Brown, issued a formal apology, on behalf of "all those who live freely thanks to Alan's work," for the "inhumane" treatment Turing received. "We're sorry, you deserved so much better," he said. Turing's imaginary machine did more against tyranny than von Neumann's real one ever did.

17

Smarter, Happier, More Productive

"I don't own a computer, have no idea how to work one," Woody Allen once told an interviewer. Most of us have come to find computers indispensable, but he manages to have a productive life without one. Are those of us with computers really better off?

There are two ways that computers might add to our well-being. First, they could do so indirectly, by increasing our ability to produce other goods and services. In this they have proved something of a disappointment. In the early 1970s, American businesses began to invest heavily in computer hardware and software, but for decades this enormous investment seemed to pay no dividends. As the economist Robert Solow put it in 1987, "You can see the computer age everywhere but in the productivity statistics." Perhaps too much time was wasted in training employees to use computers; perhaps the sorts of activities that computers make more efficient, like word processing, don't really add all that much to productivity; perhaps information becomes less valuable when it's more widely available. Whatever the case, it wasn't until the late 1990s that some of the productivity gains promised by the computer-driven "new economy" began to show up—in the United States, at any rate. So far, Europe appears to have missed out on them.

The other way computers could benefit us is more direct. They might make us smarter, or even happier. They promise to bring us such primary goods as pleasure, friendship, sex, and knowledge. If some lotus-eating visionaries are to be believed, computers may even have a spiritual dimension: as they grow ever more powerful, they have the potential to become our "mind children." At some point—the "singularity"—in the not-so-distant future, we humans will merge with these silicon creatures, thereby transcending our biology and achieving immortality. It is all this that Woody Allen is missing out on.

But there are also skeptics who maintain that computers are having the opposite effect on us: they are making us less happy, and perhaps even stupider. Among the first to raise this possibility was the American literary critic Sven Birkerts. In his 1994 book, *The Gutenberg Elegies*, Birkerts argued that the computer and other electronic media were destroying our capacity for "deep reading." His writing students, thanks to their digital devices, had become mere skimmers and scanners and scrollers. They couldn't lose themselves in a novel the way he could. This didn't bode well, Birkerts thought, for the future of literary culture.

Suppose we found that computers are diminishing our capacity for certain pleasures, or making us worse off in other ways. Why couldn't we simply spend less time in front of the screen and more time doing the things we used to do before computers came along—like burying our noses in novels? Well, it may be that computers are affecting us in a more insidious fashion than we realize. They may be reshaping our brains—and not for the better. That was the drift of "Is Google Making Us Stupid?," a 2008 cover story by Nicholas Carr in *The Atlantic*. A couple of years later, Carr, a technology writer and former executive editor of the *Harvard Business Review*, elaborated his indictment of digital culture into a book, *The Shallows: What the Internet Is Doing to Our Brains*.

Carr believes that he was himself an unwitting victim of the computer's mind-altering powers. Now in late middle age, he describes his life as a two-act play: "Analogue Youth" followed by "Digital Adulthood." In 1986, five years out of college, he dismayed his wife by spending nearly all their savings on an early version of the Apple Mac. Soon afterward, he says, he lost the ability to edit or revise on paper. Around 1990, he acquired a modem and an AOL subscription, which entitled him to spend five hours a week online sending e-mail, visiting chat rooms, and reading old newspaper articles. It was around this time that the programmer Tim Berners-Lee wrote the code for the

World Wide Web, which, in due course, Carr would be restlessly exploring with the aid of his new Netscape browser. "You know the rest of the story because it's probably your story too," he tells us. "Ever-faster chips. Ever-quicker modems. DVDs and DVD burners. Gigabyte-sized hard drives. Yahoo and Amazon and eBay. MP3s. Streaming video. Broadband. Napster and Google. BlackBerrys and iPods. Wi-Fi networks. YouTube and Wikipedia. Blogging and microblogging. Smartphones, thumb drives, netbooks. Who could resist? Certainly not I."

It wasn't until 2007, Carr says, that he had a great epiphany: "The very way my brain worked seemed to be changing." Lest we take him to be speaking metaphorically, Carr launches into a brief history of brain science, which culminates in a discussion of "neuroplasticity": the idea that experience affects the structure of the brain. Scientific orthodoxy used to hold that the adult brain was fixed and immutable: experience could alter the strengths of the connections among its neurons, it was believed, but not its overall architecture. By the late 1960s, however, striking evidence of brain plasticity began to emerge. In one series of experiments, researchers cut nerves in the hands of monkeys, and then, using microelectrode probes, observed that the monkeys' brains reorganized themselves to compensate for the peripheral damage. Later, tests on people who had lost an arm or a leg revealed something similar: the brain areas that used to receive sensory input from the lost limbs seemed to get taken over by circuits that register sensations from other parts of the body (which may account for the "phantom limb" phenomenon). Signs of brain plasticity have been observed in healthy people, too. Violinists, for instance, tend to have larger cortical areas devoted to processing signals from their fingering hands than do non-violinists. And brain scans of London cabdrivers taken in the 1990s revealed that they had larger-than-normal posterior hippocampi—structures in a part of the brain that stores spatial representations—and that the increase in size was proportional to the number of years the drivers had been in the job.

The brain's ability to change its own structure, as Carr sees it, is nothing less than "a loophole for free thought and free will." But, he hastens to add, "bad habits can be ingrained in our neurons as easily as good ones." Indeed, neuroplasticity has been invoked to explain depression, tinnitus, pornography addiction, and masochistic self-mutilation (this last is supposedly a result of pain pathways getting rewired to the brain's pleasure centers). Once new neural circuits become established in our brains, they demand to be fed, and they can hijack brain areas devoted to valuable mental skills. Thus, Carr writes, "the possibility of intellectual decay is inherent in the malleability of our brains." And the Internet "delivers precisely the kind of sensory and cognitive stimuli—repetitive, intensive, interactive, addictive—that have been shown to result in strong and rapid alterations in brain circuits and functions." He quotes the brain scientist Michael Merzenich, a pioneer of neuroplasticity and the man behind the monkey experiments in the 1960s, to the effect that the brain can be "massively remodeled" by exposure to the Internet and online tools like Google. "THEIR HEAVY USE HAS NEUROLOGICAL CONSEQUENCES," Merzenich warns in caps—in a blog post, no less.

Many in the neuroscience community scoff at such claims. The brain is not "a blob of clay pounded into shape by experience," Steven Pinker has insisted. Its wiring may change a bit when we learn a new fact or skill, but its basic cognitive architecture remains the same. And where is the evidence that using the Internet can "massively remodel" the brain? The only germane study that Carr is able to cite was undertaken in 2008 by Gary Small, a professor of psychiatry at UCLA. Small recruited a dozen experienced web surfers and a dozen novices and scanned their brains while they did Google searches. Sure enough, the two groups showed different patterns of neural firing. The activity was broader in the experienced web surfers; in particular, they made heavy use of the dorsolateral prefrontal cortex, an area of the brain associated with decision making and problem solving. In the novices, by contrast, this area was largely quiet.

Is "broader" the same as "worse"? Rather the opposite, one might think. As Carr admits, "The good news here is that Web surfing, because it engages so many brain functions, may help keep older people's minds sharp." Nor did the brain changes caused by web surfing seem to interfere with reading. When the researchers had the subjects read straight texts, there was no significant difference in brain activity between the computer veterans and the novices. And just how extensive was the alleged rewiring? When the UCLA researchers had the novices spend an hour a day surfing the web, it took only five days for their brain patterns to look like those of the veterans. "Five hours on the Internet, and the naïve subjects had already rewired their brains," Small concluded. Typically, though, brain changes that occur very quickly can also be reversed very quickly. If, for example, a normally sighted person is made to wear a blindfold, in a week's time the visual centers of his brain will have been taken over to a significant degree by the tactile centers. (This was found in experiments on the learning of Braille.) But it takes only a single day after the blindfold is removed for brain function to snap back to normal.

If surfing the web stimulates problem-solving and decision-making areas of the brain, as the UCLA study indicated, are we entitled to conclude, *pace* Carr, that Google makes us smarter? That depends on what you mean by "smart." Psychologists distinguish two broad types of intelligence. "Fluid" intelligence is one's ability to solve abstract problems, like logic puzzles. "Crystallized" intelligence is one's store of information about the world, including learned shortcuts for making inferences about it. (As one might guess, fluid intelligence tends to decline with age, while the crystallized variety tends to increase, up to a point.)

There is plenty of evidence that computers can stoke fluid intelligence. Ever played a video game? Maybe you should have. Video gamers are better at paying attention to several things at once than non-players and are better at ignoring irrelevant features of a problem. Very young children trained with video games have been shown to develop superior attention-management skills, scoring substantially higher than their untrained peers on some IQ tests. You can actually see the improvement on an EEG: four-year-olds trained on video games display patterns of activity in the attention-control parts of their brains that you'd normally expect to find in six-year-olds.

Carr acknowledges the evidence that video games can enhance certain cognitive skills. But he insists that these skills "tend to involve lower-level, or more primitive, mental functions." Those who are unfamiliar with video games might find that plausible, but a very different picture emerges from Steven Johnson's 2005 book, *Everything Bad Is Good for You*. According to Johnson, sophisticated video games (unlike the simplistic *Pac-Man*-style games of yesteryear) involve richly imagined worlds with their own hidden laws. To navigate such worlds, one must constantly frame and test hypotheses about their underlying logic. This is hardly a pastime that promotes mental flightiness. "The average video game takes about forty hours to play," Johnson observes, "the complexity of the puzzles and objectives growing steadily over time as the game progresses."

Even if computers can improve our fluid intelligence, perhaps they are inimical to crystallized intelligence—that is, to the acquisition of knowledge. This seems to be Carr's fallback position. "The Net is making us smarter," he writes, "only if we define intelligence by the Net's own standards. If we take a broader and more traditional view of intelligence—if we think about the depth of our thought rather than just its speed—we have to come to a different and considerably darker conclusion." Why is the "buzzing" brain of the computer user inferior to the "calm mind" of the book reader? Because, Carr submits, a buzzing brain is an overloaded one. Our ability to acquire knowledge depends on information getting from our "working memory"—the mind's temporary scratch pad—into our long-term memory. The working memory contains what we are conscious of at a given moment; it is estimated that it can hold only as many as four items of information at a time, which quickly vanish if they are not refreshed. The working memory is thus the bottleneck in the learning process—or, to use Carr's image, the "thimble" by which we must fill up the "bathtub" of our long-term memory. A book provides a "steady drip" of information that through sustained concentration we can transfer by means of this thimble with little spillage. But on the web, Carr writes, "we face many information faucets, all going full blast. Our little thimble overflows as we rush from one faucet to the next," and what we end up with is "a jumble of drops from different faucets, not a continuous, coherent stream from one source."

This is a seductive model, but the empirical support for Carr's conclusion is both slim and equivocal. To begin with, there is evidence that web surfing can increase the capacity of working memory. And while some studies have indeed shown that "hypertexts" impede retention—in a 2001 Canadian study, for instance, people who read a version of Elizabeth Bowen's story "The Demon Lover" festooned with clickable links took longer and reported more confusion about the plot than did those who read it in an old-fashioned "linear" text—others have failed to substantiate this claim. No study has shown that Internet use degrades the ability to learn from a book, though that doesn't stop people from feeling that this is so; one medical blogger quoted by Carr laments, "I can't read *War and Peace* anymore."

The digerati are not impressed by such avowals. "No one reads *War and Peace*," responds Clay Shirky, a digital-media scholar at New York University. "The reading public has increasingly decided that Tolstoy's sacred work isn't actually worth the time it takes to read it." (Woody Allen solved that problem by taking a speed-reading course and then reading *War and Peace* in one sitting. "It was about Russia," he said afterward.) The only reason we used to read big long novels before the advent of the Internet was that we were living in an information-impoverished environment. Our "pleasure cycles" are now tied to the web, the literary critic Sam Anderson claimed in a 2009 cover story in *New York* magazine, "In Defense of Distraction." "It's too late," he declared, "to just retreat to a quieter time."

This sort of "outré posturing" by intellectuals rankles with Carr because, he thinks, it enables ordinary people "to convince themselves that surfing the Web is a suitable, even superior, substitute for deep reading and other forms of calm and attentive thought." But Carr doesn't do enough to dissuade us from this conclusion. He fails to clinch his case that the computer is making us stupider. Can he convince us that it is making us less happy?

Suppose, like good Aristotelians, we equate happiness with human flourishing. One model for human flourishing is the pastoral ideal of quiet contemplation. It is this ideal, Carr submits, that is epitomized by "the pensive stillness of deep reading." He gives us a brisk history of reading from the invention of the codex to the Gutenberg revolution and describes how its evolution gave rise to an "intellectual ethic"—a set of normative assumptions about how the human mind works. "To read a book was to practice an unnatural process of thought, one that demanded sustained, unbroken attention to a single, static object," he writes. As written culture superseded oral culture, chains of reasoning became longer and more complex but also clearer. Library architecture came to accommodate the novel habit of reading silently to oneself as private carrels and cloisters were torn out and replaced with grand public rooms. And the miniaturization of the book, hastened in 1501 when the Italian printer Aldus Manutius introduced the pocket-sized octavo format, brought reading out of libraries into everyday life. "As our ancestors imbued their minds with the discipline to follow a line of argument or narrative through a succession of printed pages, they became more contemplative, reflective, and imaginative," Carr writes.

The digital world, by contrast, promotes a very different model of human flourishing: an industrial model of hedonic efficiency, in which speed trumps depth and pensive stillness gives way to a cataract of sensation. "The Net's interactivity gives us powerful new tools for finding information, expressing ourselves, and conversing with others," but it "also turns us into lab rats constantly pressing levers to get tiny pellets of social or intellectual nourishment."

So which fits better with your ideal of eudaemonia, deep reading or power browsing? Should you set up housekeeping in Sleepy Hollow or next to the information superhighway? The solution, one might decide, is to opt for a bit of both. But Carr seems to think that it's impossible to strike a balance. There is no stable equilibrium between analog and digital. "The Net commands our attention with far greater insistency than our television or radio or morning newspaper ever did," he writes. Once it has insidiously rewired our brains and altered our minds, we're as good as lost. He quotes the novelist Benjamin Kunkel on this loss of autonomy: "We don't feel as if we had freely chosen our online practices. We feel instead . . . that we are not distributing our attention as we intend or even like to."

Carr tells us of his own attempt to emancipate himself from the nervous digital world and return to a Woody Allen–like condition of contemplative calm. He and his wife move from "a highly connected suburb of Boston" to the mountains of Colorado, where there is no mobile phone reception. He cancels his Twitter account, suspends his Facebook membership, shuts down his blog, curtails his Skyping and instant messaging, and—"most important"—resets his e-mail so that it checks for new messages only once an hour instead of every minute. And, he confesses, he's "already backsliding." He can't help finding the digital world "cool," adding, "I'm not sure I could live without it."

Perhaps what he needs are better strategies of self-control. Has he considered disconnecting his modem and FedExing it to himself over-night, as some digital addicts say they have done? After all, as Steven Pinker has noted, "distraction is not a new phenomenon." Pinker scorns the notion that digital technologies pose a hazard to our intelligence or well-being. Aren't the sciences doing well in the digital age? he asks. Aren't philosophy, history, and cultural criticism flourishing too? There is a reason the new media have caught on, Pinker observes: "Knowledge is increasing exponentially; human brainpower and waking hours are not." Without the Internet, how can we possibly keep up with humanity's ballooning intellectual output?

This raises a prospect that has exhilarated many of the digerati. Perhaps the Internet can serve not merely as a supplement to mem-ory but as a replacement for it. "I've almost given up making an effort to remember anything," says Clive Thompson, a writer for *Wired*, "because I can instantly retrieve the information online." David Brooks, in his *New York Times* column, writes, "I had thought that the magic of the information age was that it allowed us to know more, but then I realized the magic of the information age is that it allows us to know less. It provides us with external cognitive servants—silicon memory systems, collaborative online filters, consumer preference algorithms and networked knowledge. We can burden these servants and liberate ourselves."

Books also serve as external memory-storage devices; that is why Socrates, in the *Phaedrus*, warned that the innovation of writing would lead to the atrophy of human memory. But books have expanded the reservoir of information and ideas and, through the practice of atten-tive reading, have enriched the memory, not superseded it. The Inter-net is different. Thanks to algorithmic search engines like Google, the whole universe of online information can be scanned in an instant. Not only do you not have to remember a fact; you don't even have to remember where to look it up. In time, even the intermediary of a com-puter screen might prove unnecessary—why not implant a wireless Google connection right in the head? "Certainly," says Sergey Brin, one of the founders of Google, "if you had all the world's information directly attached to your brain, or an artificial brain that was smarter than your brain, you'd be better off."

The idea that machine might supplant Mnemosyne is abhorrent to

Carr, and he devotes the most interesting portions of *The Shallows* to combating it. He gives a lucid account of the molecular basis of memory, and of the mechanisms by which the brain consolidates short-term memories into long-term ones. Biological memory, which is necessarily in "a perpetual state of renewal," is in no way analogous to the storage of bits of data in static locations on a hard drive: that he makes quite plain. Yet he doesn't answer the question that really concerns us: Why is it better to knock information into your head than to get it off the web?

The system by which the brain stores and retrieves memories is, for all its glorious intricacy, a bit of a mess. That's understandable: it's the product of blind evolution, not of rational engineering. Unlike a computer, which assigns each bit of information a precise address in its data banks, human memory is organized contextually. Items are tied together in complex associative webs and are retrieved by clues rather than by location. Ideally, the desired item just pops into your head ("The founder of phenomenology? Husserl!"). If not, you try various clues, which may or may not work ("The founder of phenomenology? Let's see, starts with an *h* . . . Heidegger!"). Human memory has certain advantages over computer memory; for instance, it tends to give automatic priority to the most frequently needed items. But it's fragile and unreliable. Unrehearsed memories soon sink into oblivion. And interference between items within associative webs causes confusion and leads to the formation of false memories.

The computer's postal-code memory system has no such vulnerabilities; each item is assigned a specific address in the computer's data bank, and retrieving that item simply means going to the relevant address. Moreover, as the cognitive psychologist Gary Marcus points out in his 2008 book, *Kluge*, it's possible to have the benefits of contextual memory without the costs. "The proof is Google," Marcus writes. "Search engines start with an underlying substrate of postal-code memory (the well-mapped information they can tap into) and build contextual memory on top. The postal-code foundation guarantees reliability, while the context on top hints at which memories are most likely needed at a given moment." It's a pity, Marcus adds, that evolution didn't start with a memory system more like the computer's.

Considering these advantages, why not outsource as much of our memory as possible to Google? Carr responds with a bit of rhetorical bluster. "The Web's connections are not *our* connections," he writes. "When we outsource our memory to a machine, we also outsource a very important part of our intellect and even our identity." Then he quotes William James, who in an 1892 lecture on memory declared, "The connecting *is* the thinking." And James was onto something: the role of memory in thinking, and in creativity.

What do we really know about creativity? Very little. We know that creative genius is not the same thing as intelligence. In fact, beyond a certain minimum IQ threshold—about one standard deviation above average, or an IQ of 115—there is no correlation at all between intelligence and creativity. We know that creativity is empirically correlated with mood-swing disorders. A couple of decades ago, Harvard researchers found that people showing "exceptional creativity"— which they put at less than 1 percent of the population—were more likely to suffer from manic depression or to be near-relatives of manic-depressives. As for the psychological mechanisms behind creative genius, those remain pretty much a mystery. About the only point generally agreed on is that, as Pinker put it, "geniuses are wonks." They work hard; they immerse themselves in their genre.

Could this immersion have something to do with stocking the memory? As an instructive case of creative genius, consider the French mathematician Henri Poincaré, who died in 1912. Poincaré's genius was distinctive in that it embraced nearly the whole of mathematics, from pure (number theory) to applied (celestial mechanics). Along with his German coeval David Hilbert, Poincaré was the last of the universalists. His powers of intuition enabled him to see deep connections between seemingly remote branches of mathematics. He virtually created the modern field of topology, framing the "Poincaré conjecture" for future generations to grapple with, and he beat Einstein to the mathematics of special relativity. Unlike many geniuses, Poincaré was a man of great practical prowess; as a young engineer, he conducted on-the-spot diagnoses of mining disasters. He was also a lovely prose stylist who wrote bestselling works on the philosophy of science; he is the only mathematician ever inducted into the literary section of the Institut de France.

What makes Poincaré such a compelling case is that his breakthroughs tended to come in moments of sudden illumination. One of the most remarkable of these was described in his essay "Mathematical

Creation." Poincaré had been struggling for some weeks with a deep issue in pure mathematics when he was obliged, in his capacity as mine inspector, to make a geological excursion. "The changes of travel made me forget my mathematical work," he recounted. "Having reached Coutances, we entered an omnibus to go some place or other. At the moment I put my foot on the step the idea came to me, without anything in my former thoughts seeming to have paved the way for it, that the transformations I had used to define the Fuchsian functions were identical with those of non-Euclidean geometry. I did not verify the idea; I should not have had time, as, upon taking my seat in the omnibus, I went on with a conversation already commenced, but I felt a perfect certainty. On my return to Caen, for conscience's sake, I verified the result at my leisure."

How to account for the full-blown revelation that struck Poincaré in the instant that his foot touched the step of the bus? His own guess was that it had arisen from unconscious activity in his memory. "The role of this unconscious work in mathematical invention appears to me incontestable," he wrote. "These sudden inspirations . . . never happen except after some days of voluntary effort which has appeared absolutely fruitless." The seemingly fruitless effort fills the memory banks with mathematical ideas—ideas that then become "mobilized atoms" in the unconscious, arranging and rearranging themselves in endless combinations, until finally the "most beautiful" of them makes it through a "delicate sieve" into full consciousness, where it will then be refined and proved.

Poincaré was a modest man, not least about his memory, which he called "not bad" in the essay. In fact, it was prodigious. "In retention and recall he exceeded even the fabulous Euler," one biographer declared. (Euler, the most prolific mathematician of all—the constant e takes his initial—was reputedly able to recite the *Aeneid* from memory.) Poincaré read with incredible speed, and his spatial memory was such that he could remember the exact page and line of a book where any particular statement had been made. His auditory memory was just as well developed, perhaps owing to his poor eyesight. In school, he was able to sit back and absorb lectures without taking notes despite being unable to see the blackboard.

It is the connection between memory and creativity, perhaps, that should make us most wary of the web. "As our use of the Web makes it harder for us to lock information into our biological memory, we're forced to rely more and more on the Net's capacious and easily searchable artificial memory," Carr observes. But conscious manipulation of externally stored information is not enough to yield the deepest of creative breakthroughs: this is what the example of Poincaré suggests. Human memory, unlike machine memory, is dynamic. Through some process we only crudely understand—Poincaré himself saw it as the collision and locking together of ideas into stable combinations—novel patterns are unconsciously detected, novel analogies discovered. And this is the process that Google, by seducing us into using it as a memory prosthesis, threatens to subvert.

It's not that the web is making us less intelligent; if anything, the evidence suggests it sharpens more cognitive skills than it dulls. It's not that the web is making us less happy, although there are certainly those who, like Carr, feel enslaved by its rhythms and cheated by the quality of its pleasures. It's that the web may be an enemy of creativity. Which is why Woody Allen might be wise in avoiding it altogether.

By the way, it is customary for reviewers of books like Carr's to note, in a jocular aside, that they interrupted their writing labors many times to update their Facebook page, to fire off text messages, to check their e-mail, to tweet and blog and amuse themselves on the Internet trying to find images of cats that look like Hitler. Well, I'm not on Facebook, and I don't know how to tweet. I have an e-mail account with AOL (America's Oldest Luddites), but there's rarely anything in my in-box. I've never had an iPod or a BlackBerry. I've never had a smartphone, or indeed a mobile phone of any kind. Like Woody Allen, I've avoided the snares of the digital age. And I still can't get anything done.

PART VII

The Cosmos Reconsidered

18

The String Theory Wars:
Is Beauty Truth?

It is the best of times in physics. Physicists are on the verge of obtaining the long-sought theory of everything. In a few elegant equations, perhaps concise enough to be emblazoned on a T-shirt, this theory will reveal how the universe began and how it will end. The key insight is that the smallest constituents of the world are not particles, as had been supposed since ancient times, but "strings"—tiny strands of energy. By vibrating in different ways, these strings produce the essential phenomena of nature, the way violin strings produce musical notes. String theory isn't just powerful; it's also mathematically beautiful. All that remains to be done is to write down the actual equations. This is taking a little longer than expected. But, with almost the entire theoretical-physics community working on the problem—presided over by a sage in Princeton, New Jersey—the millennia-old dream of a final theory is sure to be realized before long.

It is the worst of times in physics. For more than a generation, physicists have been chasing a will-o'-the-wisp called string theory. The beginning of this chase marked the end of what had been three-quarters of a century of progress. Dozens of string-theory conferences have been held, hundreds of new Ph.D.'s have been minted, and thousands of papers have been written. Yet, for all this activity, not a single new testable prediction has been made; not a single theoretical puzzle has been solved. In fact, there is no theory so far—just a set of hunches and calculations suggesting that a theory might exist. And, even if it does, this theory will come in such a bewildering number of versions that it will be of no practical use: a theory of nothing. Yet the physics establishment promotes string theory with irrational fervor, ruthlessly weeding dissenting physicists from the profession. Meanwhile, physics is stuck in a paradigm doomed to barrenness.

So which is it: the best of times or the worst of times? This is, after all, theoretical physics, not a Victorian novel. If you are a casual reader of science articles in the newspaper, you are probably more familiar with the optimistic view. But string theory has always had a few vocal skeptics. Almost three decades ago, Richard Feynman dismissed it as "crazy," "nonsense," and "the wrong direction" for physics. Sheldon Glashow, who was awarded a Nobel Prize for making one of the last great advances in physics before the beginning of the string-theory era, has likened string theory to a "new version of medieval theology" and campaigned to keep string theorists out of his own department at Harvard. (He failed.)

In 2006, two members of the string-theory generation came forward with exposés of what they deem the mess in theoretical physics. "The story I will tell could be read by some as a tragedy," Lee Smolin writes in *The Trouble with Physics: The Rise of String Theory, the Fall of a Science, and What Comes Next*. Peter Woit, in *Not Even Wrong: The Failure of String Theory and the Search for Unity in Physical Law*, prefers the term "disaster." Both Smolin and Woit were journeymen physicists when string theory became fashionable, in the early 1980s. Both are now outsiders: Smolin, a reformed string theorist (he wrote eighteen papers on the subject), has helped found a sort of Menshevik cell of physicists in Canada called the Perimeter Institute; Woit abandoned professional physics for mathematics (he is a lecturer in the mathematics department at Columbia), which gives him a cross-disciplinary perspective.

Each of these critics of string theory delivers a bill of indictment that is a mixture of science, philosophy, aesthetics, and, surprisingly, sociology. Physics, in their view, has been overtaken by a cutthroat culture that rewards technicians who work on officially sanctioned problems and discourages visionaries in the mold of Albert Einstein. Woit argues that string theory's lack of empirical grounding and conceptual rigor has left its practitioners unable to distinguish between a scientific hoax and a genuine contribution. Smolin adds a moral dimension to his plaint, linking string theory to the physics profession's "blatant prejudice" against women and blacks. Pondering the cult of empty mathematical virtuosity, he asks, "How many leading theoretical physicists were once insecure, small, pimply boys who got their revenge besting the jocks (who got the girls) in the one place they could—math class?"

It is strange to think that such sordid motives might affect something as pure and objective as physics. But these are strange days in the discipline. For the first time in its history, theory has caught up with experiment. In the absence of new data, physicists must steer by something other than hard empirical evidence in their quest for a final theory. And that something they call beauty. But in physics, as in the rest of life, beauty can be a slippery thing.

The gold standard for beauty in physics is Albert Einstein's general theory of relativity. What makes it beautiful? First, there is its simplicity. In a single equation, it explains the force of gravity as a curving in the geometry of space-time caused by the presence of mass: mass tells space-time how to curve; space-time tells mass how to move. Then there is its surprise: Who would have imagined that this whole theory would flow from the natural assumption that all frames of reference are equal, that the laws of physics should not change when you hop on a merry-go-round? Finally, there is its aura of inevitability. Nothing about it can be modified without destroying its logical structure. The physicist Steven Weinberg has compared it to Raphael's *Holy Family*, in which every figure on the canvas is perfectly placed and there is nothing you would have wanted the artist to do differently.

Einstein's general relativity was one of two revolutionary innovations in the early part of the twentieth century that inaugurated the modern era in physics. The other was quantum mechanics. Of the two, quantum mechanics was the more radical departure from the old Newtonian physics. Unlike general relativity, which dealt with well-defined objects existing in a smooth (albeit curved) space-time geometry, quantum mechanics described a random, choppy microworld where change happens in leaps, where particles act like waves (and vice versa), and where uncertainty reigns.

In the decades after this dual revolution, most of the action was on the quantum side. In addition to gravity, there are three basic forces that govern nature: electromagnetism, the "strong" force (which holds the nucleus of an atom together), and the "weak" force (which causes radioactive decay). Eventually, physicists managed to incorporate all three into the framework of quantum mechanics, creating the "standard model" of particle physics. The standard model is something of a stick-and-bubble-gum contraption: it clumsily joins very dissimilar kinds of interactions, and its equations contain about twenty arbitrary-seeming numbers—corresponding to the masses of the various particles, the ratios of the force strengths, and so on—that had to be experimentally measured and put in "by hand." Still, the standard model has proved to be splendidly useful, predicting the result of every subsequent experiment in particle physics with exquisite accuracy, often down to the eleventh decimal place. As Feynman once observed, that's like calculating the distance from Los Angeles to New York to within a hairbreadth.

The standard model was hammered out by the mid-1970s and has not had to be seriously revised since. (A crowning confirmation came in 2012 when the Higgs boson, the last missing piece, was discovered thanks to the Large Hadron Collider at CERN, the European center for experimental physics.) The standard model tells how nature behaves on the scale of molecules, atoms, electrons, and on down, where the force of gravity is weak enough to be overlooked. General relativity tells how nature behaves on the scale of apples, planets, galaxies, and on up, where quantum uncertainties average out and can be ignored. Between the two theories, all nature seems to be covered. But most physicists aren't happy with this division of labor. Everything in nature, after all, interacts with everything else. Shouldn't there be a single set of rules for describing it, rather than two inconsistent sets? And what happens when the domains of the two theories overlap—that is, when the very massive is also the very small? Just after the big bang, for example, the entire mass of what is now the observable universe was packed into a volume the size of an atom. At that tiny scale, quantum uncertainty causes the smooth geometry of general relativity to break up, and there is no telling how gravity will behave. To understand the birth of the universe, we need a theory that "unifies" general relativity and quantum mechanics. That is the theoretical physicist's dream.

String theory came into existence by accident. In the late 1960s, a couple of young physicists thumbing through mathematics books happened upon a centuries-old formula, the Euler beta function, that, miraculously, seemed to fit the latest experimental data about elementary particles. At first, no one had a clue why this should be. Within a few years, however, the hidden meaning of the formula emerged: if elementary particles were thought of as tiny wriggling strings, it all made sense. What were these strings supposed to be made of? Nothing, really. As one physicist put it, they were to be thought of as "tiny one-dimensional rips in the smooth fabric of space."

This wasn't the only way in which the new theory broke with previous thinking. We seem to live in a world that has three spatial dimensions (along with one time dimension). But for string theory to make mathematical sense, the world must have nine spatial dimensions. Why don't we notice the six extra dimensions? Because, according to string theory, they are curled up into some micro-geometry that makes them invisible. (Think of a garden hose: from a distance it looks one-dimensional, like a line; up close, however, it can be seen to have a second dimension, curled up into a little circle.) The assumption of hidden dimensions struck some physicists as extravagant. To others, though, it seemed a small price to pay. In Smolin's words, "String theory promised what no other theory had before—a quantum theory of gravity that is also a genuine unification of forces and matter."

But when would it make good on that promise? In the decades since its possibilities were first glimpsed, string theory has been through a couple of "revolutions." The first took place in 1984, when some potentially fatal kinks in the theory were worked out. On the heels of this achievement, four physicists at Princeton, dubbed the Princeton String Quartet, showed that string theory could indeed encompass all the forces of nature. Within a few years, physicists around the world had written more than a thousand papers on string theory. The theory also attracted the interest of the leading figure in the world of theoretical physics, Edward Witten.

Witten, now at the Institute for Advanced Study in Princeton, is held in awe by his fellow physicists, who have been known to compare him to Einstein. As a teenager, he was more interested in politics than in physics. In 1968, at the age of seventeen, he published an article in *The Nation* arguing that the New Left had no political strategy. He majored in history at Brandeis and worked on George McGovern's 1972 presidential campaign. (McGovern wrote him a letter of recommendation for graduate school.) When he decided to pursue a career in physics, he proved to be a quick study: Princeton Ph.D., Harvard postdoc, full professorship at Princeton at the age of twenty-nine, MacArthur "genius grant" two years later. Witten's papers are models of depth and clarity. Other physicists attack problems by doing complicated calculations; he solves them by reasoning from first principles. Witten once said that "the greatest intellectual thrill of my life" was learning that string theory could encompass both gravity and quantum mechanics. His string-theoretic investigations have led to stunning advances in pure mathematics, especially in the abstract study of knots. In 1990, he became the first physicist to be awarded the Fields Medal, considered the Nobel Prize of mathematics.

It was Witten who ushered in the second string-theory revolution, which addressed a conundrum that had arisen, in part, from all those extra dimensions. They had to be curled up so that they were invisibly small, but it turned out that there were various ways of doing this, and physicists were continually finding new ones. If there was more than one version of string theory, how could we decide which version was correct? No experiment could resolve the matter, because string theory concerns energies far beyond those that can be attained by particle accelerators. By the early 1990s, no fewer than five versions of string theory had been devised. Discouragement was in the air. But the mood improved markedly when, in 1995, Witten announced to an audience of string theorists at a conference in Los Angeles that these five seemingly distinct theories were mere facets of something deeper, which he called "M-theory." In addition to vibrating strings, M-theory allowed for vibrating membranes and blobs. As for the name of the new theory, Witten was noncommittal; he said that "M stands for magic, mystery, or membrane, according to taste." Later, he mentioned "murky" as a possibility, because "our understanding of the theory is, in fact, so primitive." Other physicists have suggested "matrix," "mother" (as in "mother of all theories"), and "masturbation." The skeptical Sheldon Glashow wondered whether the *M* wasn't an upside-down *W*, for "Witten."

Today, more than two decades after the second revolution, the theory formerly known as strings remains a seductive conjecture rather than an actual set of equations, and the nonuniqueness problem has grown to ridiculous proportions. At the latest count, the number of string theories is estimated to be something like one followed by five hundred zeros. "Why not just take this situation as a *reductio ad absurdum*?" Smolin asks. But some string theorists are unabashed: each member of this vast ensemble of alternative theories, they observe, describes a different possible universe, one with its own "local weather" and history. What if all these possible universes actually exist? Perhaps every one of them bubbled into being just as our universe did. (Physicists who believe in such a "multiverse" sometimes picture it as a cosmic champagne glass frothing with universe-bubbles.) Most of these universes will not be bio-friendly, but a few will have precisely the right conditions for the emergence of intelligent life-forms like us. The fact that our universe appears to be fine-tuned to engender life is not a matter of luck. Rather, it is a consequence of the "anthropic principle": if our universe weren't the way it is, we wouldn't be here to observe it. Partisans of the anthropic principle say that it can be used to weed out all the versions of string theory that are incompatible with our existence, and so rescue string theory from the problem of nonuniqueness.

Copernicus might have dislodged man from the center of the universe, but the anthropic principle seems to restore him to that privileged position. Many physicists despise it; one has depicted it as a "virus" infecting the minds of his fellow theorists. Others, including Witten, accept the anthropic principle, but provisionally and in a spirit of gloom. Still others seem to take perverse pleasure in it. The controversy among these factions has been likened by one participant to "a high-school-cafeteria food fight."

In their books against string theory, Smolin and Woit view the anthropic approach as a betrayal of science. Both agree with Karl Popper's dictum that if a theory is to be scientific, it must be open to falsification. But string theory, Woit points out, is like Alice's Restaurant, where, as Arlo Guthrie's song had it, "you can get anything you want." It comes in so many versions that it predicts anything and everything. In that sense, string theory is, in the words of Woit's title, "not even wrong." Supporters of the anthropic principle, for their part, rail against the "Popperazzi" and insist that it would be silly for physicists to reject string theory because of what some philosopher said that science should be. Steven Weinberg, who has a good claim to be the father of the standard model of particle physics, has argued that anthropic reasoning may open a new epoch. "Most advances in the history of science have been marked by discoveries about nature," he has observed, "but at certain turning points we have made discoveries about science itself."

Is physics, then, going postmodern? (At Harvard, as Smolin notes, the string-theory seminar was for a time actually called "Postmodern Physics.") The modern era of particle physics was empirical; theory developed in concert with experiment. The standard model may be ugly, but it works, so presumably it is at least an approximation of the truth. In the postmodern era, we are told, aesthetics must take over where experiment leaves off. Because string theory does not deign to be tested directly, its beauty must be the warrant of its truth.

In the past century, physicists who have followed their aesthetic sense in the absence of experimental data seem to have done quite well. As Paul Dirac said, "Anyone who appreciates the fundamental harmony connecting the way Nature runs and general mathematical principles must feel that a theory with the beauty and elegance of Einstein's theory has to be substantially correct."

The idea that "beauty is truth, truth beauty" may be a beautiful one, but is there any reason to think it is true? Truth, after all, is a relationship between a theory and the world, whereas beauty is a relationship between a theory and the mind. Perhaps, some have conjectured, a kind of cultural Darwinism has drilled it into us to take aesthetic pleasure in theories that are more likely to be true. Or perhaps physicists are somehow inclined to choose problems that have beautiful solutions rather than messy ones. Or perhaps nature itself, at its most fundamental level, possesses an abstract beauty that a true theory is bound to mirror. What makes all these explanations suspect is that standards of theoretical beauty tend to be ephemeral, routinely getting overthrown in scientific revolutions. "Every property that has at some date been seen as aesthetically attractive in theories has at other times been judged as displeasing or aesthetically neutral," James W. McAllister, a philosopher of science, has observed.

The closest thing to an enduring mark of beauty is simplicity; Pythagoras and Euclid prized it, and contemporary physicists continue to pay lip service to it. All else being equal, the fewer the equations, the greater the elegance. And how does string theory do by this criterion? Pretty darn well, one of its partisans has facetiously observed, since the number of defining equations it has so far produced remains precisely zero. At first, string theory seemed the very Tao of simplicity, reducing all known particles and forces to the notes of a vibrating string. As one of its pioneers commented, "String theory was too beautiful a mathematical structure to be completely irrelevant to nature." Over the years, though, it has repeatedly had to be jury-rigged in the face of new difficulties, so that it has become a Rube Goldberg machine—or, rather, a vast landscape of them. Its proponents now inveigh against what they call "the myth of uniqueness and elegance." Nature is not simple, they maintain, nor should our ultimate theory of it be. "A good, honest look at the real world does not suggest a pattern of mathematical minimality," says the Stanford physicist Leonard Susskind, who seems to have no regrets about string theory's having "gone from being Beauty to the Beast."

If neither predictive value nor beauty explains the persistence of string theory, then what does? Since the late eighteenth century, no major scientific theory has been around for more than a decade without getting a thumbs-up or a thumbs-down. Correct theories nearly always triumph quickly. But string theory, in one form or another, has been hanging on inconclusively for almost half a century now. Einstein's own pursuit of a unified theory of physics in the last three decades of his life is often cited as a case study in futility. Have a thousand string theorists done any better?

The usual excuse offered for sticking with what increasingly looks like a failed program is that no one has come up with any better ideas for unifying physics. But critics of string theory like Smolin and Woit have a different explanation, one that can be summed up in the word "sociology." They worry that academic physics has become dangerously like what the social constructivists have long charged it with being: a community that is no more rational or objective than any other group of humans. In today's hypercompetitive environment, the best hope for a young theoretical physicist is to curry favor by solving a set problem in string theory. "Nowadays," one established figure in the field has said, "if you're a hot-shot young string theorist you've got it made."

Some detect a cultlike aspect to the string-theory community, with

Witten as the guru. Smolin deplores what he considers the shoddy scientific standards that prevail in the string-theory community, where long-standing but unproved conjectures are assumed to be true because "no sensible person"—that is, no member of the tribe—doubts them. The most hilarious symptom of string theory's lack of rigor is the so-called Bogdanov affair, in which the French twin brothers Igor and Grichka Bogdanov managed to publish egregiously nonsensical articles on string theory in five peer-reviewed physics journals. Was it a reverse Sokal hoax? (In 1996, the physicist Alan Sokal fooled the editors of the postmodern journal *Social Text* into publishing an artful bit of drivel on the "hermeneutics of quantum gravity.") The Bogdanov brothers indignantly denied it, but even the Harvard string-theory group was said to be unsure, alternating between laughter at the obviousness of the fraud and hesitant concession that the authors might have been sincere.

Let's assume that the situation in theoretical physics is as bad as critics like Smolin and Woit say it is. What are non-physicists supposed to do about it? Should we form a sort of children's crusade to capture the holy land of physics from the string-theory usurpers? And whom should we install in their place?

The current problem with physics, according to Smolin, is basically a problem of style. The initiators of the dual revolution a century ago—Einstein, Bohr, Schrödinger, Heisenberg—were deep thinkers, or "seers." They confronted questions about space, time, and matter in a philosophical way. The new theories they created were essentially correct. But "the development of these theories required a lot of hard technical work, and so for several generations physics was 'normal science' and was dominated by master craftspeople," Smolin observes. "The paradoxical situation of string theory—so much promise, so little fulfillment—is exactly what you get when a lot of highly trained master craftspeople try to do the work of seers." Today, the challenge of unifying physics will require another revolution, one that mere virtuoso calculators are ill-equipped to carry out. And the solution, presumably, is to cultivate a new generation of seers.

"How strange it would be if the final theory were to be discovered in our own lifetimes!" Steven Weinberg once observed. Such a discovery, Weinberg added, would mark the sharpest discontinuity in intellectual history since the beginning of modern science, in the seventeenth century. Of course, it is possible that a final theory will never be found, that neither string theory nor any of the alternatives championed by string theory's opponents will come to anything. Perhaps the most fundamental truth about nature is simply beyond the human intellect, the way that quantum mechanics is beyond the intellect of a dog. Or perhaps, as Karl Popper believed, there will prove to be no end to the succession of deeper and deeper theories. And, even if a final theory is found, it will leave the questions about nature that most concern us—how the brain gives rise to consciousness, how we are constituted by our genes—untouched. Theoretical physics will be finished, but the rest of science will hardly notice.

19

Einstein, "Spooky Action," and the Reality of Space

In physics, as in politics, there is a time-honored notion that all action is ultimately local. Aptly enough, physicists call this the principle of locality. What the principle of locality says, in essence, is that the world consists of separately existing physical objects and that these objects can directly affect one another only if they come into contact.

It follows from the principle of locality that remote things can affect each other only indirectly, through causal intermediaries that bridge the distance between them. I can affect you, for instance, by extending my arm and giving you a pat on the cheek, or by calling you on your cell phone (electromagnetic radiation), or even—very, very slightly—by wiggling my little finger (gravitational waves). But I can't affect you in a way that jumps instantly across the expanse of space that separates us, without anything traveling from me to you—by sticking a pin in a voodoo doll, say. That would be a "nonlocal" influence.

The idea of locality emerged early in the history of science. For the Greek atomists, it was what distinguished naturalistic explanations from magical ones. Whereas the gods were believed to be capable of acting nonlocally, by simply willing remote events to occur, genuine causality for the atomists was always local, a matter of hard little atoms bumping into one another. Aristotle adhered to the principle of locality; so did Descartes. Newton (to his own distress) seemed to depart from it, because gravity in his theory was an attractive force that somehow reached across empty space, perhaps instantaneously. But in the nineteenth century, Michael Faraday restored locality by introducing the concept of a field as an all-pervading, energy-carrying medium through which forces like gravity and electromagnetism are transmitted from one object to another—not instantaneously, as would be the case with nonlocal action, but at a fixed and finite speed: the speed of light.

The principle of locality promises to render the workings of nature rational and transparent, allowing complex phenomena to be "reduced" to local interactions. Nonlocality, by contrast, has always been the refuge of the occult and the hermetic, of believers in "sympathies" and "synchronicity" and "holism."

Albert Einstein had a deep philosophical faith in the principle of locality. He couldn't imagine how science could proceed without it. "Unless one makes this kind of assumption," Einstein said, "physical thinking in the familiar sense would not be possible." He dismissed the possibility of voodoo-like, space-defying, nonlocal influences as "spooky action at a distance" (*spukhafte Fernwirkung*).

But in the 1920s, Einstein, alone among his contemporaries, noticed something disturbing: the new science of quantum mechanics looked to be at odds with the principle of locality. It seemed to entail "spooky action at a distance." He took this to mean that there must be something seriously amiss with the quantum theory, which he himself had helped create. (Einstein's 1921 Nobel Prize was for his work on the photoelectric effect, a quantum phenomenon, not for his discovery of relativity.) He came up with clever thought experiments to make the problem he saw vivid. Defenders of the quantum consensus, chief among them Niels Bohr, endeavored to rebut Einstein, yet they failed to appreciate the true force of his logic. Meanwhile, quantum theory's growing record of success in explaining chemical bonding and predicting new particles made Einstein's qualms look merely "philosophical"— which can be a term of abuse in physics.

And so the matter stood until 1964, a little under a decade after Einstein's death. That was when an Irish physicist named John Stewart Bell did what no one had imagined was possible: he showed that Einstein's philosophical objection could be put to an experimental test. If quantum mechanics was right, Bell proved, "spooky action" could actually be observed in the lab. And when the experiment Bell described was carried out—imperfectly in Berkeley in the 1970s, more decisively in Paris in 1982, and near authoritatively in Delft in 2015 (with still further tests to come in future years)—the "spooky" predictions of quantum mechanics were vindicated.

Yet the reaction to this news—from physicists with an interest in philosophy, from philosophers with an interest in physics—has been strangely equivocal. Some have declared the revelation that nature flouts the principle of locality to be "mind-boggling" (the physicist Brian Greene) and "the single most astonishing discovery of twentieth-century physics" (the philosopher Tim Maudlin). Others think that nonlocality, though perhaps a little spooky on the face of it, is nothing to get metaphysically exercised over, since it "still follows the ordinary laws of cause and effect" (the physicist Lawrence Krauss). Still others—notwithstanding Bell and the subsequent experiments—deny that the world genuinely contains nonlocal connections. Prominent among them is the Nobel laureate Murray Gell-Mann, who insists that all the talk of "action at a distance" amounts to a "flurry of flapdoodle."

There is nothing financial or personal about this disagreement over nonlocality. It is (to quote the science writer George Musser) "intellectually pure." And if it seems stubbornly unresolvable, that may be because it goes to a deeper issue: Just what should we expect from physics—a recipe for making predictions or a unified picture of reality?

That was the issue that divided Einstein and Bohr in the early days of quantum mechanics. Einstein was, metaphysically speaking, a "realist": he believed in an objective physical world, one that existed independently of our observations. And he thought the job of physics was to give a complete and intelligible account of that world. "Reality is the real business of physics" was how he put it.

Bohr, by contrast, was notoriously slippery in his metaphysical commitments. At times he sounded like an "idealist" (in the philosophical sense), arguing that physical properties become definite only when they are measured, and hence that reality is, to some extent, created by the very act of observation. At other times he sounded like an "instrumentalist," arguing that quantum mechanics was meant to be an instrument for predicting our observations, not a true representation of a world lurking behind those observations. "There is no quantum world," Bohr provocatively declared.

Bohr was happy with the quantum theory; Einstein was not. Popular accounts often claim that Einstein objected to quantum mechanics because it made randomness a fundamental ingredient of reality. "God does not play dice," he famously said. But it was not randomness per se that bothered Einstein. Rather, it was his suspicion that the appearance of randomness in quantum mechanics was a sign that the new theory didn't tell the whole story of what was going on in the physical world. And the principle of locality had an important part in this suspicion.

Here is the simplest of the thought experiments designed to illustrate Einstein's misgivings about quantum mechanics. It has become known as Einstein's Boxes, since it was originally presented by Einstein in 1927 (although it was later reformulated by de Broglie, Schrödinger, and Heisenberg). Start with a box that contains a single particle—say, an electron. According to quantum mechanics, an electron confined to a box does not have a definite location until we look inside to see just where it is. Prior to that act of observation, the electron is in a mixture of potential locations spread throughout the box. This mixture is mathematically represented by a "wave function," which expresses the different probabilities of detecting the electron at the various locations inside the box if you do an experiment. (In French, the wave function is evocatively called *densité de présence*.) Only when the observation is made does potentiality turn into actuality. Then the wave function "collapses" (as physicists say) to a single point, and the electron's location becomes definite.

Now suppose that before any such observational experiment is carried out, we put a partition through the middle of the box containing the electron. If this is done in the appropriate way, the wave function of the electron inside will be split in two: loosely speaking, half of the wave function will be to the left of the partition, half to the right. This is a complete quantum description of the physical situation: there is no deeper fact about which side of the partition the electron is "really" on. The wave function does not represent our ignorance of where the particle is; it represents genuine indeterminacy in the world.

Next, we separate the two partitioned halves of the box. We put the left half box on a plane for Paris, and the right half box on a plane for Tokyo. Once the boxes have arrived at their respective destinations, a physicist in Tokyo does an experiment to see whether there is an electron in the right half box. Quantum mechanics says the result of this experiment will be purely random, like flipping a coin. Because the wave function is equally split between the two half boxes, there is a fifty-fifty chance that the Tokyo physicist will detect the presence of an electron.

Well, suppose she does. With that, the wave function collapses. The act of detecting an electron in the Tokyo box causes the part of the wave function associated with the Paris box to vanish instantaneously. It's as though the Paris box telepathically knows the (supposedly random) outcome of the Tokyo experiment and reacts accordingly. Now if a physicist in Paris looks in the left half box, he is certain not to find an electron. (Of course, the "collapse" could have gone the other way, and the Paris physicist could have found the electron.)

That is how things are supposed to work according to the orthodox quantum story, as developed by Bohr, Heisenberg, and other founders of the theory. It is called the Copenhagen interpretation of quantum mechanics, since Bohr was the head of the physics institute at the University of Copenhagen. According to the Copenhagen interpretation, the very act of observation causes the spread-out probability wave to collapse into a sharply located particle. Hence what has been called the best explanation of quantum mechanics in five words or fewer: "Don't look: waves. Look: particles."

To Einstein, this was absurd. How could merely looking inside a box cause spread-out potentiality to snap into sharp actuality? And how could looking inside a box in Tokyo instantly change the physical state of a box on the other side of the world in Paris? That would be "spooky action at a distance"—a clear contravention of the principle of locality. So something must be wrong with the Copenhagen interpretation.

Einstein's intuition was just what common sense would suggest: the particle must have been in one box or the other all along. Therefore, Einstein concluded, quantum mechanics must be incomplete. It offers a blurry picture of a sharp reality rather than (as defenders of the Copenhagen interpretation insisted) a sharp picture of a blurry reality.

Bohr never confronted the simple logic of Einstein's Boxes. Instead, he focused his polemical attention on a later and more elaborate thought experiment, one that Einstein came up with in the 1930s after he had left Germany and relocated to the Institute for Advanced Study in Princeton. It is referred to by the initials "EPR," after Einstein and his two junior collaborators, Boris Podolsky (from Russia) and Nathan Rosen (from Brooklyn).

The EPR thought experiment involves a pair of particles that get created together and then go their separate ways. Einstein saw that according to quantum mechanics such particles would be "entangled": they would stay correlated in how they responded to experiments regardless of how far apart they moved. As an example, consider what happens when an "excited" atom—an atom whose energy level has been artificially boosted—sheds its excess energy by emitting a pair of photons (particles that are components of light). These two photons will fly off in contrary directions; eventually, they will travel to opposite sides of the galaxy and beyond. Yet quantum mechanics says that no matter how vast the separation between them the two photons will remain entangled as a single quantum system. When subjected to the same experiment, each will respond exactly as its partner does. If, for example, you see the near photon successfully make its way through a polarizing filter (like the kind in sunglasses), you automatically know that its distant partner would do so as well, provided the near and distant filters are set at the same angle.

You might think that such entangled particles are no more mysterious than a pair of identical twins who have moved to different cities; if you see that twin A in New York has red hair, you automatically know that twin B in Sydney is a redhead too. But unlike hair color, quantum properties do not become definite until they are subjected to a measurement. When particle A is measured, it snaps out of a mixture of possibilities into a definite state, and this supposedly forces its entangled partner, B, to snap out of its own mixture of possibilities into an exactly correlated state.

If quantum mechanics is right, the entangled particles are not like a pair of identical twins; rather, they are like the magical coins that are sometimes imagined in thought experiments: although they are not altered or weighted in any manner, they know always to land the same way when flipped. It is as though there were a telepathic link between the entangled particles, one that enables them to coordinate their behavior instantaneously across vast distances—even though all known methods of communication are, in accord with relativity, limited by the speed of light.

Einstein's conclusion in the EPR thought experiment was the same as in Einstein's Boxes: such a link would be "spooky action at a distance." Quantum entanglement can't be real. The tightly choreographed behavior of the widely separated particles must be preprogrammed from the start (as with identical twins), not a matter of correlated randomness (as with magical coins). And because quantum theory doesn't account for such preprogramming—referred to by physicists as hidden variables—it gives an incomplete description of the world.

The EPR reasoning is clear enough up to this point. But the paper that Einstein, Podolsky, and Rosen published in 1935 went further, attempting to discredit the Heisenberg uncertainty principle, which states that certain pairs of physical properties of a particle—such as the particle's position and its momentum—cannot both be definite at the same time. (Einstein later blamed this overreach on the younger Podolsky, who wrote the last section of the EPR paper.) That muddied matters sufficiently to give Bohr the opening he needed for his rebuttal, which proved to be a masterpiece of obscurity. A decade after he produced it, Bohr confessed that he himself had difficulty making sense of what he had written. Yet most physicists, tired as they were of this "philosophical" dispute and eager to return to their quantum calculations, simply assumed that Bohr had won the debate over the has-been Einstein. According to Einstein's biographer Abraham Pais, his "fame would be undiminished, if not enhanced, if he had gone fishing instead."

One later physicist who stood apart from this consensus was John Stewart Bell (1928–1990). The son of a Belfast horse trader, Bell made his career in applied physics, helping to design the first particle accelerator at CERN (the European physics center near Geneva). But he also looked on the conceptual foundations of physics with a philosopher's eye. In the clarity and rigor of his thought, Bell rivaled Einstein. And like Einstein, he had misgivings about quantum mechanics. "I hesitated to think it might be wrong, but I knew that it was rotten," he said.

Reflecting on the EPR thought experiment, Bell ingeniously saw a way to tweak it so that it could be made into a *real* experiment, one that would force the issue between quantum mechanics and locality. His proof that this was possible, now famous as "Bell's theorem," was published in 1964. Amazingly, it required just a couple of pages of high school algebra.

The gist of Bell's idea was to get entangled particles to reveal their nonlocal connection—if indeed there was one—by interrogating them more subtly. This could be done, he saw, by measuring the spin of the particles along different angles. Because of the peculiarities of quantum spin, each measurement would be like asking the particle a yes-or-no question. If two separated but entangled particles are asked the same question—that is, if their spins are measured along the same angle—they are guaranteed to give the same answer: either both yes or both no. There's nothing necessarily magical about such agreement: it could have been programmed into the pair of entangled particles when they were created together.

But if entangled particles are asked *different* questions—that is, if their respective spins are measured along different angles—quantum mechanics then predicts a precise statistical pattern of matches and mismatches in their yes-or-no answers. And with the right combination of questions, Bell proved, the pattern predicted by quantum mechanics would be unambiguously nonlocal. No amount of pre-programming, no "hidden variables" of the kind envisaged by Einstein, could explain it. Such a tight correlation, Bell demonstrated, could only mean that the separated particles were coordinating their behavior in some way not yet understood—that each "knew" not only which question its distant twin was being asked but also how the twin answered.

So what Bell did was this. First, he conceived an experiment in which a certain combination of measurements would be made on a pair of separated but entangled particles. Then he showed, by an ironclad mathematical argument, that *if* the statistical pattern arising from these measurements was the one predicted by quantum mechanics, *then* there would be no escape, logically speaking, from spooky action.

All that remained to settle Einstein's quarrel with quantum mechanics was to do the experiment Bell outlined and see whether this statistical pattern emerged. It took technology a little while to catch up, but by the early 1970s physicists had begun to test Bell's idea in the lab. In experiments measuring properties of pairs of entangled photons, the pattern of statistical correlation Bell identified has invariably been observed. The verdict: Spooky action is real.*

So was Einstein wrong? It would be fairer (if a bit melodramatic) to say that he was betrayed by nature—which, by violating the principle of locality, turned out to be less reasonable than he imagined. Yet Einstein had seen more deeply into quantum mechanics than Bohr and the other defenders of quantum orthodoxy. (Einstein once remarked that he had given a hundred times as much thought to quantum mechanics as he had to his own theory of relativity.) He realized that nonlocality was a genuine and disturbing feature of the new theory and not, as Bohr and his circle seemed to regard it, a mere mathematical fiction.

Let's pause here to note just how strange the quantum connection between entangled particles really is. First, it is undiluted by distance—unlike gravity, which falls off in strength. Second, it is discriminating: an experiment done on one photon in an entangled pair affects only its partner, wherever that partner may be, leaving all other photons, near and far, untouched. The discriminating nature of entanglement again stands in contrast to gravity, where a disturbance created by the jostling of one atom will ripple out to affect every atom in the universe. And third, the quantum connection is instantaneous: a change in the state of one entangled particle makes itself felt on its partner without delay, no matter how vast the gulf that separates them—yet again in contrast to gravity, whose influence travels at the speed of light.

It is the third of these features of quantum nonlocality, its instantaneousness, that is the most worrisome. As Einstein realized early, it would mean that the entangled particles were communicating faster than the speed of light, which is generally forbidden by the theory of relativity. If, for example, particle A is near the earth and its entangled twin, B, is near Alpha Centauri (the nearest star system to the sun), a measurement performed on A will alter the state of B instantly, even though it would take 4.3 years for light to get from A to B.

---

*Diehards have tried to hold on to locality by invoking loopholes like "superdeterminism" or "backward causation," but it is hard to imagine Einstein buying any of these extravagant metaphysical hypotheses.

Many physicists tend to brush off this apparent conflict between relativity theory and quantum mechanics. They point out that even though quantum entanglement does seem to entail "superluminal" (faster than light) influences, those influences can't be used for communication—to send messages, say, or music. There is no possibility of a "Bell telephone" (as in John, not Alexander Graham). And the reason is quantum randomness: although entangled particles do exchange information between themselves, a would-be human signaler can't control their random behavior and encode a message in it. Since it can't be used for communication, quantum entanglement doesn't give rise to the sorts of causal anomalies Einstein warned about—like being able to send a message backward in time. So quantum theory and relativity, though conceptually at odds with each other, can "peacefully coexist."

For John Bell, that wasn't good enough. "We have an apparent incompatibility, at the deepest level, between the two fundamental pillars of contemporary theory," he observed in a 1984 lecture. If our picture of physical reality is to be coherent, Bell believed, the tension between relativity theory and quantum mechanics must be confronted.

In 2006, an impressive breakthrough along these lines was made by Roderich Tumulka, a German-born mathematician at Rutgers. Building on the insights of Bell and other philosophically minded physicists, Tumulka succeeded in creating a model of nonlocal entanglement that fully abides by Einsteinian relativity. Contrary to what is widely believed, relativity does not completely rule out influences that are faster than light. (Indeed, physicists sometimes talk about hypothetical particles called tachyons that move faster than the speed of light.) What relativity *does* rule out is absolute time: a universal "now" that is valid for all observers. Entangled particles would seem to require such a universal clock if they are to synchronize their behavior across vast distances. But Tumulka found an ingenious (though exceedingly subtle) way around this. He showed how a certain speculative extension of quantum mechanics—known, for complicated reasons, as flashy GRW—could allow entangled particles to act in synchrony without violating relativity's ban on absolute simultaneity. Although the mechanism behind this nonlocal "spooky action" remains obscure, Tumulka at least proved that it is logically consistent with relativity after all—a result that might well have surprised Einstein.

However it works, nonlocality has subversive implications for our understanding of space. Its discovery suggests that we might live in a "holistic" universe, one in which things that seem to be far apart may, at a deeper level of reality, not be truly separate at all. The space of our everyday experience might be an illusion, a mere projection of some more basic causal system. A nice metaphor for this (proposed by the philosopher Jenann Ismael) is the kaleidoscope. Don't think of entangled particles as "magical coins" somehow exchanging messages across space. Rather, think of them as being like the multiple images of a glass bead tumbling about in a kaleidoscope—different mirror reflections of the same underlying particle.

Despite such radical implications, the physics profession has (for the most part) taken the demonstration of nonlocality in stride. Younger physicists who have grown up with nonlocality don't find it all that spooky. "The kids here say, that's just the way it is," observes the experimental physicist Nicolas Gisin. Among the elder generation, there seems to be a widespread impression that the weirdness of nonlocality can be evaded by taking a "non-realist" view of quantum mechanics— by looking upon it the way Niels Bohr did, as a mathematical device for making predictions, not a picture of reality. One contemporary representative of this way of thinking is Stephen Hawking, who has said, "I don't demand that a theory correspond to reality because I don't know what it is . . . All I'm concerned with is that the theory should predict the results of measurements."

Yet a deeper understanding of entanglement and nonlocality is also crucial to resolving the perennial argument over how to "interpret" quantum mechanics—how to give a realistic account of what happens when a measurement is made and the wave function mysteriously and randomly "collapses." This is the very problem that vexed Einstein, and it is one that still vexes a small and contentious community of physicists (like Sir Roger Penrose, Sheldon Goldstein, and Sean Carroll) and philosophers of physics (like David Z. Albert, Tim Maudlin, and David Wallace) who continue to demand from physics the same thing that Einstein did: a unified and intelligible account of how the world really is. For them, the conceptual foundations of quantum mechanics, and the role of "spooky action" in those foundations, remain very much a work in progress.

20

How Will the Universe End?

One of my favorite moments in Woody Allen's film *Annie Hall* is when Alvy Singer (Allen's alter ego) is shown having an existential crisis as a little boy. His mother summons a psychiatrist, one Dr. Flicker, to find out what's wrong.

"Why are you depressed, Alvy?" Dr. Flicker asks.

"The universe is expanding," Alvy says. "The universe is everything, and if it's expanding, someday it will break apart and that will be the end of everything."

"Why is that your business?" interrupts his mother. Turning to the psychiatrist, she announces, "He's stopped doing his homework!"

"What's the point?" Alvy says.

"What has the universe got to do with it!" his mother shouts. "You're here in Brooklyn! Brooklyn is not expanding!"

Dr. Flicker jumps in: "It won't be expanding for billions of years yet, Alvy, and we've got to try and enjoy ourselves while we're here, eh? Ha ha ha." (Cut to a view of the Singer house, which happens to be under the Coney Island roller coaster.)

I used to take Dr. Flicker's side in this matter. How silly to despond about the end of everything! After all, the cosmos was born only around fourteen billion years ago, when the big bang happened, and parts of it will remain hospitable to our descendants for a good hundred billion years, even as the whole thing continues to spread out.

A couple of decades ago, however, astronomers peering through their telescopes began to notice something rather alarming. The expansion of the universe, their observations indicated, was not proceeding at the stately, ever-slowing pace that Einstein's equations had predicted. Instead, it was speeding up. Some "dark energy" was evidently pushing against gravity, sending galaxies hurtling away from one another at a runaway rate. New measurements after the turn of the millennium confirmed this strange finding. On July 22, 2003, *The New York Times* ran an ominous headline: "Astronomers Report Evidence of 'Dark Energy' Splitting the Universe." David Letterman found this so disturbing that he mentioned it several consecutive nights in his *Late Show* monologue, wondering why the *Times* buried the story on page A13.

Until recently, the ultimate destiny of the universe looked a little more hopeful—or remote. Back around the middle of the last century, cosmologists figured out that there were two possible fates for the universe. Either it would continue to expand forever, getting very cold and very dark as the stars winked out one by one, the black holes evaporated, and all material structures disintegrated into an increasingly dilute sea of elementary particles: the big chill. Or it would eventually stop expanding and collapse back upon itself in a fiery, all-annihilating implosion: the big crunch.

Which of these two scenarios would come to pass depended on one crucial thing: how much stuff there was in the universe. So, at least, said Einstein's general theory of relativity. Stuff—matter and energy—creates gravity. And, as every undergraduate physics major will tell you, gravity sucks. It tends to draw things together. With enough stuff, and hence enough gravity, the expansion of the universe would eventually be arrested and reversed. With too little stuff, the gravity would merely slow the expansion, which would go on forever. So, to determine how the universe would ultimately expire, cosmologists thought that all they had to do was to weigh it. And preliminary estimates—taking account of the visible galaxies, the so-called dark matter, and even the possible mass of the little neutrinos that swarm through it all—suggested that the universe had only enough weight to slow the expansion, not to turn it around.

Now, as cosmic fates go, the big chill might not seem a whole lot better than the big crunch. In the first, the temperature goes to absolute zero; in the second, it goes to infinity. Extinction by fire or by ice—what's to choose? Yet a few imaginative scientists, haunted, like Woody Allen, by visions of the end of the universe, came up with formulations of how our distant descendants might manage to go on enjoying life forever, despite these unpleasant conditions. In the big chill scenario, our descendants could have an infinity of slower and slower experiences, with lots of sleep in between. In the big crunch scenario, they could have an infinity of faster and faster experiences in the run-up to the final implosion. Either way, the progress of civilization would be unlimited. No cause for existential gloom.

But the dark-energy news seemed to change all that. (No wonder David Letterman was disturbed by it.) It spells inescapable doom for intelligent life in the far, far future. No matter where you are located, the rest of the universe would eventually be receding from you at the speed of light, slipping forever beyond the horizon of knowability. Meanwhile, the shrinking region of space still accessible to you would be filling up with a kind of insidious radiation that would eventually choke off information processing—and with it the very possibility of thought. We seem to be headed not for a big crunch or a big chill but for something far nastier: a big crack-up. "All our knowledge, civilization and culture are destined to be forgotten," one prominent cosmologist declared to the press. It looks as if little Alvy Singer was right after all. The universe is going to "break apart," and that will indeed mean the end of everything—even Brooklyn.

Hearing this news made me think of the inscription that someone once said should be on all churches: "Important if true." Applied to cosmology—the study of the universe as a whole—that is a big if. Cosmic speculations that make it into the newspapers should often be taken with a pinch of salt. Back in the 1990s, some astronomers from Johns Hopkins made headlines by announcing that the cosmos was *turquoise*; just two months later, they made headlines again by announcing that, no, it was actually *beige*. This may be a frivolous example, but even in graver matters—like the fate of the universe—cosmologists tend to reverse themselves every decade or so. As one of them once told me, cosmology is not really a science at all, because you can't do experiments with the universe. It's more like a detective story. Even the term that is sometimes applied to theorizing about the end of the universe, "eschatology" (from the Greek word for "farthest"), is borrowed from theology.

Before I was going to start worrying about the extinction of absolutely everything in some inconceivably distant epoch, I thought it would be a good idea to talk to a few leading cosmologists. Just how certain were they that the cosmos was undergoing a disastrous runaway expansion? Was intelligent life really doomed to perish as a result? How could they, as scientists, talk about the ultimate future of "civilization" and "consciousness" with a straight face?

It seemed natural to start with Freeman Dyson, an English-born physicist who has been at the Institute for Advanced Study in Princeton since the 1940s. Dyson is one of the founding fathers of cosmic eschatology, which he concedes is a "faintly disreputable" subject. He is also a fierce optimist about the far future, one who envisions "a universe growing without limit in richness and complexity, a universe of life surviving forever and making itself known to its neighbors across unimaginable gulfs of space and time." In 1979, he wrote a paper called "Time Without End," in which he used the laws of physics to show how humanity could flourish eternally in a slowly expanding universe, even as the stars died and the chill became absolute. The trick is to match your metabolism to the falling temperature, thinking your thoughts ever more slowly and hibernating for longer and longer periods while extraneous information is dumped into the void as waste heat. In this way, Dyson calculated, a complex society could go on perpetually with a finite energy reserve, one equivalent to a mere eight hours of sunlight.

The day I went to see Dyson, it was raining in Princeton. It took me half an hour to walk from the train station to the Institute for Advanced Study, which sits by a pond in five hundred acres of woods. The institute is a serene, otherworldly place. There are no students to distract the eminent scientists and scholars in residence from pursuing their intellectual fancies. Dyson's office is in the same building where Einstein spent the last decades of his career fruitlessly searching for a unified theory of physics.

An elfin, courtly man with deep-set eyes and a hawk-like nose, Dyson frequently lapsed into silence or emitted snuffles of amusement. I started by asking him whether the evidence that the universe was caught up in an accelerated expansion had blighted his hopes for the future of civilization.

"Not necessarily," he said. "It's a completely open question whether this acceleration will continue forever or whether it will peter out after a while. There are several theories of what kind of cosmic field might be causing it and no observations to determine which of them is right. If it's caused by the so-called dark energy of empty space, then the expansion will keep speeding up forever, which is bad news as far as life is concerned. But if it's caused by some other kind of force field— which, out of ignorance, we label 'quintessence'—then the expansion might well slow down as we go into the future. Some quintessence theories even say that the universe will eventually stop expanding altogether and collapse. Of course, that, too, would be unfortunate for civilization, since nothing would survive the big crunch."

Well, then, I said, let's stick with the optimistic scenario. Suppose the acceleration does turn out to be temporary and the future universe settles into a nice cruise-control expansion. What could our descendants possibly look like a trillion trillion trillion years from now, when the stars have disappeared and the universe is dark and freezing and so diffuse that it's practically empty? What will they be made of?

"The most plausible answer," Dyson said, "is that conscious life will take the form of interstellar dust clouds." He was alluding to the kinds of inorganic life-forms imagined by the late astronomer Sir Fred Hoyle in his 1957 science-fiction novel, *The Black Cloud*. "An ever-expanding network of charged dust particles, communicating by electromagnetic forces, has all the complexity necessary for thinking an infinite number of novel thoughts."

But, I objected, can we really imagine such a wispy thing, spread out over billions of light-years of space, being conscious?

"Well," he said, "how do you imagine a couple of kilograms of protoplasm in someone's skull being conscious? We have no idea how that works either."

Practically next door to Dyson at the institute is the office of Ed Witten, a gangly fellow, now in his late sixties, who is widely regarded as the smartest physicist of his generation, if not the living incarnation of Einstein. Witten is one of the prime movers behind string theory— which, if its hairy math is ever sorted out, holds out some hope of furnishing the theory of everything that physicists have long been after. He has an unnerving ability to shuffle complicated equations in his head without ever writing anything down, and he speaks in a hushed, soft, rather high-pitched voice. Witten had been quoted in the press as calling the discovery of the runaway expansion of the universe "an extremely uncomfortable result." Why, I wondered, did he see it that way? Was it simply inconvenient for theoretical reasons? Or did he worry about its implications for the destiny of the cosmos? When I asked him, he agonized for a moment before responding, "Both."

Yet Witten, too, thought there was a good chance that the runaway expansion would be only temporary, as some of the quintessence theories predicted, rather than permanent, as the dark-energy hypothesis implied. "The quintessence theories are nicer, and I hope they're right," he told me. If the acceleration does indeed relax to zero, and the big crack-up is averted, could civilization go on forever? Witten was unsure. One cause for concern was the possibility that protons will eventually decay, resulting in the dissolution of all matter within another, oh, 10^{33} years or so. Freeman Dyson had scoffed at this when I talked with him, pointing out that no one had ever observed a proton decaying, but he insisted that intelligent beings could persist even if atoms fell to pieces, by re-embodying themselves in "plasma clouds"—swarms of electrons and positrons. I mentioned this to Witten. "Did Dyson really say that?" he exclaimed. "Good. Because I think protons probably do decay."

■

Back at the Princeton railroad station after visiting Ed Witten and Freeman Dyson, waiting for the train to New York and munching on a vile "veggie" sandwich that I had picked up at the convenience store across the parking lot, I pondered proton decay and Dyson's scenario for eternal life. How would his sentient black clouds, be they made up of cosmic dust or of electron-positron plasma, while away the aeons in an utterly freezing and dark universe? What passions would engross their infinite number of ever-slowing thoughts? After all (as Alvy Singer's alter ego once observed), eternity is a long time, especially toward the end. Maybe they would play games of cosmic chess, in which each move took trillions of years. But even at that rate they would run through every possible game of chess in a mere $10^{10^{70}}$ years—long before the final decay of the burned-out cinders of the stars. What then? Would they come around to George Bernard Shaw's conclusion (reached by him at the age of ninety-two) that the prospect of personal immortality was an "unimaginable horror"? Or would they feel that, subjectively at least, time was passing quickly enough?

It was almost with a sense of relief that I spoke to Lawrence Krauss a few days later. Krauss, a boyish fellow even in his early sixties who now directs the Origins Project at Arizona State University, is one of the physicists who guessed on purely theoretical grounds, even before the astronomical data came in, that the cosmos might be undergoing a runaway expansion. "We appear to be living in the worst of all possible universes," Krauss told me, clearly relishing the note of anti-Leibnizian pessimism he struck. "If the runaway expansion keeps going, our knowledge will actually decrease as time passes. The rest of the universe will be literally disappearing before our very eyes surprisingly soon—in the next ten or twenty billion years. And life is doomed—even Freeman Dyson accepts that. But the good news is that we can't prove we're living in the worst of all possible universes. No finite set of data will ever enable us to predict the fate of the cosmos with certainty. And, in fact, that doesn't really matter. Because, unlike Freeman, I think that we're doomed even if the runaway phase turns out to be only temporary."

What about Dyson's vision of a civilization of sentient dust clouds living forever in an expanding universe, entertaining an infinite number of thoughts on a finite store of energy? "It turns out, basically for mathematical reasons, that there's no way you can have an infinite number of thoughts unless you do a lot of hibernating," Krauss said. "You sleep for longer and longer periods, waking up for short intervals to think—sort of like an old physicist. But what's going to wake you up? I have a teenage daughter, and I know that if I didn't wake her up, she'd sleep forever. The black cloud would need an alarm clock that would wake it up an infinite number of times on a finite amount of energy. When a colleague and I pointed this out, Dyson came up with a cute alarm clock that could actually do this, but then we argued that this alarm clock would eventually fall apart because of quantum mechanics."

So, regardless of the fate of the cosmos, things look pretty hopeless for intelligent life in the long run. But one should remember, Krauss said, that the long run is a very long time. He told me about a meeting he once attended at the Vatican on the future of the universe: "There were about fifteen people, theologians, a few cosmologists, some biologists. The idea was to find common ground, but after three days it was clear that we had nothing to say to one another. When theologians talk about the 'long term,' raising questions about resurrection and such, they're really thinking about the short term. We weren't even on the same plane. When you talk about 10^{50} years, the theologians' eyes glaze over. I told them that it was important that they listen to what I had to say; theology, if it's relevant, has to be consistent with science. At the same time I was thinking, 'It doesn't matter what you have to say, because whatever theology has to say is irrelevant to science.'"

At least one cosmologist I knew of would be quite happy to absorb theology into physics, especially when it came to talking about the end of the universe. That's Frank Tipler, a professor at Tulane University in New Orleans. In 1994, Tipler published a strangely ingenious book called *The Physics of Immortality*, in which he argued that the big crunch would be the happiest possible ending for the cosmos. The final moments before universal annihilation would release an infinite amount of energy, Tipler reasoned, and that could drive an infinite amount of computation, which would produce an infinite number of thoughts—a subjective eternity. Everyone who ever existed would be "resurrected" in an orgy of virtual reality, which would correspond pretty neatly to what religious believers have in mind when they talk about heaven. Thus, while the physical cosmos would come to an abrupt end in the big crunch, the mental cosmos would go on forever.

Was Tipler's blissful eschatological scenario—which he called the Omega Point—spoiled by the news that the cosmos seemed to be caught up in a runaway expansion? He certainly didn't think so when I talked to him. "The universe has no choice but to expand to a maximum size and then contract to a final singularity," he exclaimed in his thick southern drawl. (He's a native of Alabama and a self-described "redneck.") Any other cosmic finale, he said, would violate a certain law of quantum mechanics called unitarity. Moreover, "the known laws of physics require that intelligent life persist until the end of time and gain control of the universe."

When I mentioned that Freeman Dyson (among others) could not see why this should be so, Tipler shouted in exasperation, "Ah went up to Princeton last November and ah *tode* him the argument! Ah *tode* him!"

Then he told me too. It was long and complicated, but the nub of it was that intelligent beings must be present at the end to sort of massage the big crunch in a certain way so that it would not violate another law of quantum mechanics, the Bekenstein bound. So our eternal survival is built into the very logic of the cosmos. "If the laws of PHEES-ics are with us," he roared, "who can be against us?"

Tipler's idea of an infinite frolic just before the big crunch was seductive to me—more so, at least, than Dyson's vision of a community of increasingly dilute black clouds staving off the cold in an eternal big chill. But if the universe is in a runaway expansion, both are pipe dreams. The only way to survive in the long run is to get the hell out. Yet how do you escape a dying universe if—as little Alvy Singer pointed out—the universe is everything?

A man who claims to see an answer to this question is Michio Kaku. A theoretical physicist at City College in New York, Kaku looks and talks a bit like Mr. Sulu on *Star Trek*. He is not the least bit worried about the fate of this universe. "If your ship is sinking," he said to me, "why not get a lifeboat and leave?"

We earthlings can't do this just yet, Kaku observed. That is because we are a mere type 1 civilization, able to marshal the energy only of a single planet. But eventually, assuming a reasonable rate of economic growth and technological progress, we will graduate to being a type 2 civilization, commanding the energy of a star, and thence to being a type 3 civilization, able to summon the energy of an entire galaxy. Then space-time itself will be our plaything. We'll have the power to open up a "wormhole" through which we can slip into a brand-new universe.

"Of course," Kaku added, "it may take as long as 100,000 years for such a type 3 civilization to develop, but the universe won't start getting really cold for trillions of years."

There is one other thing that the beings in such a civilization will need, Kaku stressed to me: a unified theory of physics, one that would show them how to stabilize the wormhole so it doesn't disappear before they can make their escape. The closest thing we have to that now, string theory, is so difficult that no one, not even Ed Witten, knows how to get it to work. Kaku wasn't the least bit gloomy that the universe might be dying. "In fact," he said, "I'm in a state of exhilaration, because this would force us, really force us, to crack string theory. People say, 'What has string theory done for me lately? Has it given me better cable TV reception?' What I tell them is that string theory—or whatever the final, unified theory of physics turns out to be—could be our one and only hope for surviving the death of this universe."

Although other cosmologists were rudely dismissive of Kaku's lifeboat scenario—"a good prop for a science-fiction story," said one; "somewhat more fantastical than most of *Star Trek*," remarked another—it sounded good to me. But then I started thinking. To become a type 3 civilization, one powerful enough to engineer a stable wormhole leading to a new universe, we would have to gain control of our entire galaxy. That means colonizing something like a billion habitable planets. But if this is what the future is going to look like, then almost all the intelligent observers who will ever exist will live in one of these billion colonies. So, how come we find ourselves sitting on the home planet at the very beginning of the process? The odds against being in such an unusual situation—the very earliest people, the equivalent of Adam and Eve—are a billion to one.

■

My vague qualms about the unlikeliness of Kaku's lifeboat theory were considerably sharpened when I talked to J. Richard Gott III, an astrophysicist at Princeton University. Gott is known for making bold quantitative predictions about the longevity of things—from Broadway shows like *Cats* to America's space program to intelligent life in the universe. He bases these predictions on what he calls the Copernican principle, which says, in essence, *you're not special.* "If life in the universe is going to last a long time, why do we find ourselves living when we do, only 14 billion years after the beginning?" Gott said to me, speaking in an improbable Tennessee accent whose register occasionally leaped up an octave, like Don Knotts's. "And it is a disturbing fact that we as a species have only been around for 200,000 years. If there are going to be many intelligent species descended from us flourishing in epochs far in the future, then why are we so lucky to be the first?" Doing a quick back-of-the-envelope calculation, Gott determined that it was 95 percent likely that humanity would last more than 5,100 years but would die out before 7.8 million years (a longevity that, coincidentally, is quite similar to that of other mammal species, which tend to go extinct around 2 million years after appearing). Gott was not inclined to speculate on what might do us in—biological warfare? asteroid collision? nearby supernova? sheer boredom with existence? But he did leave me feeling that the runaway expansion of our universe, if real, was the least of our worries.

Despite the pessimistic tenor of Gott's line of thought, he was positively chirpy in conversation. In fact, all the cosmologists I had spoken to so far had a certain mirthfulness about them when discussing eschatological matters—even Lawrence Krauss, the one who talked about this being the worst of all possible universes. (" 'Eschatology'— it's a great word," Krauss said. "I had never heard of it until I discovered I was doing it.") Was no one made melancholy or irritable by the prospect of our universe decaying into nothingness? I thought of Steven Weinberg, the Nobel laureate in physics who, in his 1977 book about the birth of the universe, *The First Three Minutes*, glumly observed, "The more the universe seems comprehensible, the more it also seems pointless." It was Weinberg's pessimistic conclusion in that book—he wrote that civilization faced cosmic extinction from either endless cold or unbearable heat—that had inspired Freeman Dyson to come up with his scenario for eternal life in an expanding cosmos.

I called Weinberg at the University of Texas, where he teaches. "So, you want to hear what old grumpy has to say, eh?" he growled in a deep voice. He began with a dazzling theoretical exposition that led up to a point I had heard before: no one really knows what's causing the current runaway expansion or whether it will continue forever. The most natural assumption, he added, was that it would. But he wasn't really worried about the existential implications. "For me and you and everyone else around today, the universe will be over in less than 10^2 years," he said. In his peculiarly sardonic way, Weinberg seemed as jolly as all the other cosmologists. "The universe will come to an end," he said, "and that may be tragic, but it also provides its fill of comedy. Postmodernists and social constructivists, Republicans and socialists and clergymen of all creeds—they're all an endless source of amusement."

■

It was time to tally up the eschatological results. The cosmos has three possible fates: big crunch (eventual collapse), big chill (expansion forever at a steady rate), or big crack-up (expansion forever at an accelerating rate). Humanity, too, has three possible fates: eternal flourishing, endless stagnation, or ultimate extinction. And judging from all the distinguished cosmologists who weighed in with opinions, every combination from column A and column B was theoretically open. We could flourish eternally in virtual reality at the big crunch or as expanding black clouds in the big chill. We could escape the big crunch/chill/crack-up by wormholing our way into a fresh universe. We could face ultimate extinction by being incinerated by the big crunch or by being isolated and choked off by the big crack-up. We could be doomed to endless stagnation—thinking the same patterns of thoughts over and over again, or perhaps sleeping forever because of a faulty alarm clock—in the big chill. One distinguished physicist I spoke to, Andrei Linde of Stanford University, even said that we could not rule out the possibility of there being something *after* the big crunch. For all the fascinating theories and scenarios they spin out, practitioners of cosmic eschatology are in a position very much like that of Hollywood studio heads: nobody knows anything.

Still, little Alvy Singer is in good company in being soul-sick over the fate of the cosmos, however vaguely it is descried. At the end of the nineteenth century, figures like Swinburne and Henry Adams expressed similar anguish at what then seemed to be the certain heat death of the universe from entropy. In 1903, Bertrand Russell described his "unyielding despair" at the thought that "all the labors of the ages, all the devotion, all the inspiration, all the noonday brightness of human genius, are destined to extinction in the vast death of the solar system, and that the whole temple of Man's achievement must inevitably be buried beneath the debris of a universe in ruins." Yet a few decades later, he declared such effusions of cosmic angst to be "nonsense," perhaps an effect of "bad digestion."

Why should we want the universe to last forever, anyway? Look— either the universe has a purpose or it doesn't. If it doesn't, then it is absurd. If it does have a purpose, then there are two possibilities: either this purpose is eventually achieved, or it is never achieved. If it is never achieved, then the universe is futile. But if it is eventually achieved, then any further existence of the universe is pointless. So, no matter how you slice it, an eternal universe is either (a) absurd, (b) futile, or (c) eventually pointless.

Despite this cast-iron logic, some thinkers believe that the longer the universe goes on, the better it is, ethically speaking. As John Leslie, a cosmological philosopher at the University of Guelph in Canada, told me, "This is true simply on utilitarian grounds: the more intelligent happy beings in the future, the merrier." Philosophers of a more pessimistic kidney, like Schopenhauer, have taken precisely the opposite view: life is, on the whole, so miserable that a cold and dead universe is preferable to one teeming with conscious beings.

If the current runaway expansion of the cosmos really does portend that our infinitesimal flicker of civilization will be followed by an eternity of bleak emptiness, then that shouldn't make life now any less worth living, should it? It may be true that nothing we do today will matter when the burned-out cinder of our sun is finally swallowed by a galactic black hole in a trillion trillion years. But by parity of reason, nothing that will happen in a trillion trillion years should matter to us today. In particular (as the philosopher Thomas Nagel has observed), it does not matter now that in a trillion trillion years nothing we do now will matter.

Then what is the point of cosmology? It's not going to cure cancer or solve our energy problems or give us a better sex life, obviously enough. Still, we ought to be excited that we're living in the first generation in the history of humanity that might be able to answer the question, how will the universe end? "It amazes me," Lawrence Krauss said, "that, sitting in a place on the edge of nowhere in a not especially interesting time in the history of the universe, we can, on the basis of simple laws of physics, draw conclusions about the future of life and the cosmos. That's something we should relish, regardless of whether we're here for a long time or not."

So, remember the advice offered by Monty Python in their classic "Galaxy Song." When life gets you down, the song says, and you're feeling very small and insecure, turn your mind to the cosmic sublimity of the ever-expanding universe—" 'cause there's bugger all down here on Earth."

PART VIII

Quick Studies: A Selection of Shorter Essays

Little Big Man

Is life absurd? Many people believe so, and the reasons they give often have to do with space and time. Compared with the vast universe, we are but infinitesimal specks, they say, and the human life span constitutes the merest blip on the cosmic timescale.

Others fail to see how our spatiotemporal dimensions alone could make life absurd. The philosopher Thomas Nagel, for one, has argued that if life is absurd given our present size and longevity, it would be no less absurd if we lived for millions of years or if we were big enough to fill the cosmos.

The issue of life's absurdity is moot, I suppose, but there is an interesting question lurking in the background: Which, from the point of view of the universe, is more contemptible—our minuteness or our brevity? Cosmically speaking, do we last a long time for our size or a short time? Or, put the other way, are we big or small for our life span?

The best way to go about answering this question is to look for a fundamental unit of space and of time that would render the two dimensions comparable. Here is where contemporary physics comes in handy. In trying to blend the theories that describe the very large (Einstein's general relativity) and the very small (quantum mechanics), physicists have found that it is natural to regard space and time as being composed, on the tiniest scales, of discrete quanta—geometric atoms, as it were. The shortest spatial span that has any meaning is the Planck length, which is about 10^{-35} meters (about 20 orders of magnitude smaller than a proton). The shortest possible tick of an imaginary clock (sometimes called a chronon) is the Planck time, about 10^{-43} seconds. (This is the time it takes light to cross a distance equal to the Planck length.)

Now, suppose we construct two cosmic scales, one for size and one for longevity. The size scale will extend from the smallest possible size, the Planck length, to the largest possible size, the radius of the observable universe. The longevity scale will extend from the briefest possible life span, the Planck time, to the longest possible life span, the current age of the universe.

Where do we rank on these two scales? On the cosmic size scale, humans, at a meter or two in length, are more or less in the middle. Roughly speaking, the observable universe dwarfs us the way we dwarf the Planck length. On the longevity scale, by contrast, we are very close to the top. The number of Planck times that make up a human lifetime is very, very much more than the number of human lifetimes that make up the age of the universe. "People talk about the ephemeral nature of existence," the physicist Roger Penrose has commented, "but [on such a scale] it can be seen that we are not ephemeral at all—we live more or less as long as the Universe itself!"

Certainly, then, we humans have little reason to feel angst about our temporal finitude. *Sub specie aeternitatis*, we endure for an awfully long time. But our extreme puniness certainly gives us cause for cosmic embarrassment.

Or does it? In Voltaire's philosophical tale "Micromégas," a giant from the star Sirius visits the planet Earth, where, with the aid of a magnifying instrument, he eventually detects a ship full of humans in the Baltic Sea. He is at first amazed to discover that these "invisible insects," created in the "abyss of the infinitely small," seem to possess souls. Then he wonders whether their diminutiveness might not indeed be a mark of superiority. "O intelligent atoms," he addresses them, "you must doubtless enjoy very pure pleasures on your globe, for having so little body and seeming to be all spirit, you must pass your lives in love and in thought, which is the true life of spirits." In response, the microscopic humans begin spouting philosophical inanities from Aristotle, Descartes, and Aquinas, at which the giant is overcome with laughter.

Would we humans be less absurd if we were bigger? Probably not, but we would surely be less sound. Consider a sixty-foot-tall man. (The example comes from the biologist J.B.S. Haldane's beautiful 1926 essay, "On Being the Right Size.") This giant man would be not only ten times as high as an ordinary human but also ten times as wide and thick. His total weight would therefore be a thousand times greater. Unfortunately, the cross section of his bones would only be greater by a factor of a hundred, so every square inch of his bone structure would have to support ten times the weight borne by a square inch of human bone. But the human thighbone breaks under about ten times the human weight. Consequently, when the sixty-foot man takes a step, he breaks a thigh.

If we humans are, from the cosmic perspective, absurdly tiny for our life span, perhaps we can derive some consolation from our impressive complexity of form. That is what John Donne did in his *Devotions upon Emergent Occasions*. "Man consists of more pieces, more parts, than the world," he observed. "And if those pieces were extended and stretched out in man as they are in the world, man would be the giant and the world the dwarf."

Doom Soon

Do you ever lie awake at night wondering why you happen to be alive just now? Why it should be that your own particular bit of self-consciousness popped into existence within the last few decades and not, say, during the reign of the Antonines or ten million years hence? If you do wonder about this, and if your musings take a sufficiently rigorous form, you might arrive at a terrible realization: The human race is doomed to die out—and quickly.

So, at least, a handful of cosmologists and philosophers have concluded. Their reasoning is known as the doomsday argument. It goes like this: Suppose humanity were to have a happier fate, surviving thousands or millions of years into the future. And why not? The sun still has half its ten-billion-year life span to go. The earth's population might stabilize at fifteen billion or so, and our successors could even colonize other parts of the galaxy, allowing a far greater increase in their numbers.

But think what that means: nearly every human who will ever exist will live in the distant future. This would make us unusual in the extreme. Assume, quite conservatively, that a billion new people will be born every decade until the sun burns out. That makes a total of 500 quadrillion people. At most, 50 billion people have either lived in the past or are living now. Thus we would be among the first 0.00001 percent of all members of the human species to exist. Are we really so special?

But suppose, contrariwise, that humanity will be wiped out imminently, that some sort of apocalypse is around the corner. Then it is quite reasonable, statistically speaking, that our moment is the present. After all, more than seven billion of the fifty billion humans who have ever lived are alive today, and with no future epochs to live in, this is far and away the most likely time to exist. Conclusion: doom soon.

Even as transcendental a priori arguments go, this one is pretty breathtaking. For economy of premise and extravagance of conclusion, it rivals Saint Anselm's derivation of God's existence from the idea of perfection, or Donald Davidson's proof that most of what we believe must be true or else our words would not refer to the right things.

As far as anyone knows, the doomsday argument was first publicly broached in 1983 at a meeting of the Royal Society in London. Its apparent author was Brandon Carter, an Australian astrophysicist famous for his work on black holes. A decade earlier, Carter had baptized the much-debated "anthropic principle," which purports to explain why the laws of physics look the way they do: if they were any different, life could not have emerged, and hence we would not be here to observe them. We find ourselves living in this particular universe, in other words, because alternative universes are uncongenial to intelligent life. Carter was suggesting to the Royal Society that the same goes for time: we find ourselves living in this particular epoch because earlier and later ones are, for reasons we do not fully grasp, uncongenial to us. As interesting cosmological ideas often are, the doomsday argument was soon taken up by philosophers—notably John Leslie of the University of Guelph in Ontario.

Perhaps you are skeptical about, if not plain scornful of, the doomsday argument. It looks like logical trumpery. How could an abstract argument have such an experientially rich upshot? Yet it is difficult to find anything amiss in its logic. The sole assumption it requires—an eminently plausible one—is that if humanity endures, our cumulative numbers will increase. And the inference it makes is justified by the principle of probability known as Bayes' theorem, which dictates how a piece of evidence (we are living now) should affect the likelihoods we assign to competing hypotheses (doom sooner versus doom later).

Furthermore, the doomsday argument may not seem so unlikely once you consider all the forms doom could actually take. Don't think Ebola, greenhouse, nukes; think cosmos. An asteroid might bump into our planet (one wonders whether the doomsday argument occurred to the dinosaurs sixty-five million years ago). The Swift-Tuttle comet— dubbed the Doomsday Rock by the media—will be swinging awfully close on or about August 14, 2126. And that's the small stuff. The North Star could go supernova at any moment. In fact, the dread event might already have happened, in which case the news of it is traveling earth- ward in the form of lethal radiation that will obliterate us upon receipt.

The most delicious scenario of all is the one in which absolutely ev- erything suddenly gets reduced to nothingness. Most cosmologists think the universe is far closer to the beginning of its career—the big bang— than to the end, whether that end ultimately takes the form of a big crunch or a big chill. But they cannot rule out the possibility that space is "metastable," meaning that it could spontaneously slip to a lower en- ergy level at any moment. If this were to happen, a little bubble of "true vacuum" would spontaneously appear without warning somewhere and begin to inflate at the speed of light. Its wall would contain tremen- dous energy, annihilating in a stroke everything before it: entire star systems, galaxies, galactic clusters, and eventually the cosmos itself.

Now, that's a speculative bubble worth worrying about.

Death: Bad?

To be "philosophical" about something, in common parlance, is to face it calmly, without irrational anxiety. And the paradigm of a thing to be philosophical about is death. Here Socrates is held to be the model.

Sentenced to die by an Athenian court on the charge of impiety, he serenely drank the fatal cup of hemlock. Death, he told his friends, might be annihilation, in which case it is like a long, dreamless slumber, or it might be a migration of the soul from one place to another. Either way, it is nothing to be feared.

Cicero said that to philosophize is to learn how to die—a pithy statement, but a misleading one. There is more to philosophizing than that. Broadly speaking, philosophy has three concerns: how the world hangs together (metaphysics), how our beliefs can be justified (epistemology), and how to live (ethics). Arguably, learning how to die fits under the third of these. If you wanted to get rhetorically elastic about it, you might even say that by learning how to die we learn how to live.

That thought is more or less the inspiration behind Simon Critchley's *Book of Dead Philosophers* (2008). What defines bourgeois life in the West today is our pervasive dread of death—so claims Critchley, a philosophy professor at the New School in New York. (He wrote this book, he tells us more than once, on a hill overlooking Los Angeles, which, because of "its peculiar terror of annihilation," is "surely a candidate city for the world capital of death.") As long as we are afraid of death, Critchley thinks, we cannot really be happy. And one way to overcome this fear is by looking to the example of philosophers. "I want to defend the ideal of the philosophical death," Critchley writes.

So he takes us on a breezy and often entertaining tour through the history of philosophy, looking at how 190 or so philosophers from ancient times to the present lived and died. Not all the deaths recounted are as edifying as Socrates'. Plato, for example, might have died of a lice infestation. The Enlightenment thinker La Mettrie seems to have expired after eating a quantity of truffle pâté. Several deaths are precipitated by collisions: Montaigne's brother was killed by a tennis ball; Rousseau died of cerebral bleeding, possibly as a result of being knocked down by a galloping Great Dane; and Roland Barthes was blindsided by a dry-cleaning truck after a luncheon with the politician Jack Lang. The American pragmatist John Dewey, who lived into his nineties, came to the most banal end of all: he broke his hip and then succumbed to pneumonia.

Critchley has a mischievous sense of humor, and he certainly does not shrink from the embodied nature of his subjects. There is arch merrymaking over beans (Pythagoras and Empedocles proscribed them) and flatulence (Metrocles became suicidally distraught over a bean-related gaseous indiscretion during a lecture rehearsal). We are told of Marx's genital carbuncles, Nietzsche's syphilitic coprophagy, and Freud's cancerous cheek growth, so malodorous that it repelled his favorite dog, a chow. There are Woody Allen–ish moments, as when the moribund Democritus "ordered many hot loaves of bread to be brought to his house. By applying these to his nostrils he somehow managed to postpone his death." And there are last words, the best of which belong to Heinrich Heine: "God will pardon me. It's his métier."

How are we to cultivate the wisdom necessary to confront death? It's hard to extract a consistent message from the way philosophers have died. Montaigne trained for the end by keeping death "continually present, not merely in my imagination, but in my mouth." Spinoza went to the contrary extreme, declaring, "A free man thinks least of all of death." Dying philosophically might simply mean dying cheerfully. The beau ideal is David Hume, who, when asked whether the thought of annihilation terrified him, calmly replied, "Not the least."

The idea that death is not such a bad thing may be liberating, but is it true? Ancient philosophers tended to think so, and Critchley (along with Hume) finds their attitude congenial. He writes, "The philosopher looks death in the face and has the strength to say that it is nothing."

There are three classic arguments, all derived from Epicurus and his follower Lucretius, that it is irrational to fear death. If death is annihilation, the first one goes, then there are no nasty post-death experiences to worry about. As Epicurus put it, where death is, I am not; where I am, death is not. The second says it does not matter whether you die young or old, for in either case you'll be dead for an eternity. The third points out that your nonexistence after your death is merely the mirror image of your nonexistence before your birth. Why should you be any more disturbed by the one than by the other?

Unfortunately, all three of these arguments fail to establish their conclusion. The American philosopher Thomas Nagel, in his 1970 essay "Death," showed what was wrong with the first. Just because you don't experience something as nasty, or indeed experience it at all, doesn't mean it's not bad for you. Suppose, Nagel says, an intelligent person has a brain injury that reduces him to the mental condition of a contented baby. Certainly this would be a grave misfortune for the person. Then is not the same true for death, where the loss is still more severe?

The second argument is just as poor. It implies that John Keats's demise at twenty-five was no more unfortunate than Tolstoy's at eighty-two, since both will be dead for an eternity anyway. The odd thing about this argument, as the (dead) English philosopher Bernard Williams noticed, is that it contradicts the first one. True, the amount of time you're around to enjoy the goods of life doesn't mathematically reduce the eternity of your death. But the amount of time you're dead matters only if there's something undesirable about being dead.

The third argument, that your posthumous nonexistence is no more to be feared than your prenatal nonexistence, also collapses on close inspection. As Nagel observed, there is an important asymmetry between the two abysses that temporally flank your life. The time after you die is time of which your death deprives you. You might have lived longer. But you could not possibly have existed in the time before your birth. Had you been conceived earlier than you actually were, you would have had a different genetic identity. In other words, you would not be you.

Cultivating indifference to death is not only philosophically unsound. It can be morally dangerous. If my own death is nothing, then why get worked up over the deaths of others? The barrenness of the Epicurean attitude—enjoy life from moment to moment and don't worry about death—is epitomized by George Santayana, one of Critchley's exemplary dead philosophers. After resigning from Harvard, Santayana lived in Rome, where he was discovered by American soldiers after the liberation of Italy in 1944. Asked his opinion of the war by a journalist from *Life* magazine, Santayana fatuously replied, "I know nothing; I live in the Eternal."

Contrast the example of Miguel de Unamuno, a twentieth-century Spaniard inexplicably omitted by Critchley. No one had a greater terror of death than Unamuno, who wrote, "As a child, I remained unmoved when shown the most moving pictures of hell, for even then nothing appeared to me quite so horrible as nothingness itself." In 1936, at the risk of being lynched by a Falangist mob, Unamuno publicly faced down the pro-Franco thug José Millán-Astray. Placed under house arrest, Unamuno died ten weeks later. Aptly, the Falangist battle cry Unamuno found most repellent was "¡Viva la muerte!"—Long live death!

The Looking-Glass War

One day many years ago, as I was wandering down an old tenement block on New York's Lower East Side, I chanced upon an odd little shop. It sold only one item: mirrors that do not reverse left and right, or True Mirrors, as the store calls them. There was one in the shop-window. Looking at my reflection in it, I was appalled by how crooked my facial features seemed, how lopsided my smile was, how ridiculous my hair looked parted on the wrong side of my head. Then I realized that the image I was confronting was the real me, the one the world sees. The image of myself I am used to, the one I see when I look into an ordinary mirror, is actually that of an incongruous counterpart whose left and right are the reverse of mine.

There is nothing very strange about the fact that ordinary mirrors reverse left and right, is there? "Left" and "right" are labels for the two horizontal directions parallel to the mirror. The two vertical directions parallel to the mirror are "up" and "down." But the optics and geometry of reflection are precisely the same for all dimensions parallel to the mirror. So why does a mirror treat the horizontal and vertical axes differently? Why does it reverse left and right but not up and down?

This question might seem foolish at first. "When I wave my right hand, my mirror counterpart waves his left hand," you say. "When I wiggle my head, I should scarcely expect my counterpart to wiggle his feet." True enough, but you might plausibly expect your counterpart to appear upside down, with his feet directly opposite your head—just as his left hand is directly opposite your right hand.

Foolish or not, the issue has been vexing philosophers for more than half a century now. As far as I can tell, it first arose in the early 1950s, as a sort of sidebar to discussions of Immanuel Kant's theory of spatial relations. In his 1964 book *The Ambidextrous Universe*, the science popularizer Martin Gardner stirred things up by arguing that the puzzle has a false premise. A mirror does not really reverse left and right at all, he claimed; instead, it reverses front and back along an axis perpendicular to the mirror. If you are facing north, your mirror image is facing south, but your eastward hand is directly across from his eastward hand. In Gardner's view, we merely "find it convenient" to call our image left/right reversed because we happen to be bilaterally symmetric. In 1970, the philosopher Jonathan Bennett published an article endorsing Gardner's supposed resolution of what by now had become the "mildly famous mirror problem."

But the sense of closure was premature. In 1974, the philosopher Ned Block published a long, diagram-filled piece in *The Journal of Philosophy* in which he contended that the question "Why does a mirror reverse right/left but not up/down?" has at least four different interpretations. Block claimed that the four interpretations had been clumsily conflated by Gardner and Bennett; he also insisted that in two of the four, a mirror really *does* reverse left and right. Three years later, in an equally lengthy article that appeared in *The Philosophical Review*, an English philosopher named Don Locke declared that Block was only "half right." Mirrors actually reverse left and right in every relevant sense, he argued.

Reading these papers, and others that have appeared since, one comes to feel that the mirror problem defies philosophical reflection. People can't seem to agree on the most basic facts. For example: Stand sideways to a mirror, shoulder to shoulder with your image. Your left/right axis is now perpendicular to the mirror's surface. Gardner and Bennett say that in this case and this case alone, a mirror really does reverse left and right. Block and Locke say that in this case and this case alone, left and right are the same direction for you and your mirror image. (Having just dashed into the dressing room to try this, I think I'm in the Block-Locke camp. My right arm and my mirror counterpart's right arm both pointed east; on the other hand, he was wearing his watch on his right wrist, whereas I wear mine on my left.)

The key to the mirror puzzle would seem to lie in some subtle disanalogy between left/right and up/down. Both of these pairs of directions are relative to the orientation of the body (unlike, say, east/west and skyward/earthward). But as any child will attest, left/right is much harder to master than up/down. The human body displays no gross asymmetries between its two sides. (There is, of course, the heart, but that is hidden.) So "left" and "right" have to be defined in terms of "front" and "head": your left hand is the one that is to the west when you stand on the ground and face north. This would remain true even if a surgeon cut off your two hands and sewed them onto the opposite arms.

Left/right is thus logically parasitic on front/back, whereas up/down is not. And a mirror, everyone agrees, reverses front and back. That must be why it also reverses left and right—if indeed it does, which to this day remains unclear.

Fatigued by the debate? Although the little shop I discovered on the Lower East Side is long gone, you can still get a True Mirror over the Internet. But don't try shaving in front of the thing—your face will be a bloody mess.

Astrology and the Demarcation Problem

One of the fundamental problems in the philosophy of science is called the demarcation problem: What is it that distinguishes science from nonscience or pseudoscience? What, for example, makes evolutionary theory scientific and creationism pseudoscientific?

Philosophers of science have taken three broad approaches to this problem. One approach is to look for some criterion that demarcates science from pseudoscience—like Karl Popper's criterion of falsifiability, which says that a theory is scientific if it is open to experimental refutation. Let's call this approach *methodological positivism*.

A second approach is to argue that science is demarcated from pseudoscience not by its methodology but by a sociological criterion: the judgment of the "scientific community." This view, associated with figures like Thomas Kuhn, Michael Polanyi, and Robert K. Merton, might be called *elitist authoritarianism*.

Finally, one might deny the very possibility of demarcation, arguing that there is no rationale that privileges scientific over nonscientific beliefs. This view is often called *epistemological anarchism*.

The most impish of the epistemological anarchists was Paul

Feyerabend (1924–1994), whose methodological motto was "anything goes." His good friend Imre Lakatos (1922–1974) took an opposing view—a blend, as he saw it, of Popper and Kuhn. Instead of asking whether a single theory was scientific or unscientific, Lakatos examined entire research programs, classifying them as "progressive" or "degenerating." He used the contrast to show how scientific consensus could be rational, not just a matter of mob psychology.

Feyerabend found this unpersuasive. "Neither Lakatos nor anybody else has shown that science is better than witchcraft and that science proceeds in a rational way," he wrote in a sketch for a paper titled "Theses on Anarchism." But Lakatos never gave up trying to convince Feyerabend that his views were wrongheaded, and Feyerabend returned the favor.

These friendly antagonists exchanged abundant letters on the matter, with a good deal of ribaldry—some of it of a sort that no longer evokes an easy smile. "I am very tired because my liver is acting up which is a pity, for my desire to lay the broads here (and there are some fine specimens walking around on campus) is considerably reduced," Feyerabend wrote from Berkeley. The affection between them is much in evidence. Lakatos, writing from the London School of Economics, often signed his letters "Love, Imre."

Philosophically, however, there is no detectable convergence in their positions over their years of correspondence. That is not surprising, really, given how vexed the demarcation problem is.

Take a seemingly easy case: astrology. We all think astrology is a pseudoscience (*pace* Feyerabend), but it is not easy to say why. The usual arguments are (1) astrology grew out of a magical worldview; (2) the planets are too far away for there to be any physical mechanism for their alleged influence on human character and fate; and (3) people believe in astrology only out of a desire for comforting explanations. But the first argument is also true of chemistry, medicine, and cosmology. Nor is the second decisive, for there have been many scientific theories that have lacked physical foundations. When Isaac Newton proposed his theory of gravitation, for example, he could furnish no mechanism to account for how gravity's mysterious "action at a distance" was possible. As for the third argument, people often believe in good theories for illegitimate reasons.

Surely, though, astrology fails Popper's criterion of falsifiability, doesn't it? This seems a promising line of argument, because horoscopes yield only vague tendencies, not sharp predictions. Yet such tendencies, if they exist, ought to show up as statistical correlations for large populations.

Indeed, attempts have been made to detect such correlations—notably in the 1960s by Michel Gauquelin, who surveyed the times of birth and subsequent careers of twenty-five thousand Frenchmen. Gauquelin found no significant relationship between careers and zodiac signs, which are determined by the position of the sun at the time of birth. But he did turn up associations between certain occupations of people and the positions of certain planets at the time of their birth. For instance, in accordance with the predictions of astrology, individuals born when Mars was at its zenith were more likely to become athletes, and those born when Saturn was rising were more likely to become scientists—in both cases by statistically significant margins.

But if the scientific status of astrology cannot be impugned on Popperian grounds, perhaps it can be on Lakatosian ones. Some years after Lakatos's death, the philosopher Paul R. Thagard made a detailed case for astrology being a "dramatically unprogressive" research program and hence pseudoscientific. Astrology has not added to its explanatory power since the time of Ptolemy, Thagard pointed out. It is riddled with anomalies, which the community of astrologers shows scant interest in clearing up. And it has been overtaken by alternative theories of personality and behavior, like Freudian psychology and genetics. (Not that the latter two aren't also vulnerable to the charge of being pseudoscientific.)

Lakatos himself clearly thought astrology was pseudoscience—along with much else. "The social sciences are on par with astrology, it is no use beating around the bush," he wrote to Feyerabend. ("Funny that I should be teaching at the London School of Economics!" he added.) As for Feyerabend, the only definition of science he was finally prepared to tolerate was "what follows from a principle of general hedonism." And what about the truth? "The truth, whatever it is, be damned. What we need is laughter."

Gödel Takes On the U.S. Constitution

Aristotle, the second-greatest logician of all time, was also an expert on political constitutions. Can the same be said for the greatest logician of all time, Kurt Gödel? Gödel had a genius for detecting the paradoxical in unexpected places. He looked into the axioms of mathematics and saw incompleteness. He looked into the equations of general relativity and saw "closed time-like loops." And he looked into the Constitution of the United States and saw a logical loophole that could allow a dictator to take power. Or did he?

The scene was New Jersey, 1947. Sixteen years earlier, Gödel had stunned the intellectual world by proving that no logical system could ever capture all the truths of mathematics—a result that, along with Heisenberg's uncertainty principle, was to become iconic of the limitations of human knowledge. He had left Austria for the United States when the Nazis took over, and for nearly a decade he had been a member of the Institute for Advanced Study in Princeton. It was time, he decided, to become an American citizen. Earlier that year, he had found a curious new solution to Einstein's cosmological equations, one that made space-time rotate rather than expand. In the "Gödel universe," it would be possible to journey in a big circle and arrive back at your starting point before you left.

But Gödel's research into time travel was interrupted by his citizenship hearing, scheduled for December 5 in Trenton. His character witnesses were to be his close friends Albert Einstein and the game theory co-inventor Oskar Morgenstern, who also served that day as his chauffeur. Being a fastidious man, Gödel decided to make a close study of American political institutions in preparation for the exam. On the eve of the hearing, he called Morgenstern in a state of agitation. He had found a logical inconsistency in the Constitution, he said. Morgenstern was amused by this, but he realized that Gödel was dead serious. He urged him not to mention the matter to the judge, fearing it would jeopardize his citizenship bid.

On the short drive to Trenton the next day, Einstein and Morgenstern tried to distract Gödel with jokes. When they arrived at the court, the judge, Phillip Forman, was impressed by Gödel's eminent witnesses, and he invited the trio into his chambers. After some small talk, he said to Gödel, "Up to now you have held German citizenship." No, Gödel corrected, Austrian. "Anyhow," continued the judge, "it was under an evil dictatorship . . . but fortunately that's not possible in America."

"On the contrary, I know how that can happen," cried Gödel, and he began to explain how the Constitution might permit such a thing to occur. The judge, however, indicated that he was not interested, and Einstein and Morgenstern succeeded in quieting the examinee down. (This exchange was recounted in the 1998 book *In the Light of Logic* by the logician Solomon Feferman, who apparently got it indirectly from Morgenstern.)

A few months later, Gödel took his oath of citizenship. Writing to his mother back in Vienna, he commented that "one went home with the impression that American citizenship, in contrast to most others, really meant something."

For those of us who have never read the Constitution all the way through, this anecdote cannot but be disturbing. What was the logical flaw that Gödel believed he had descried in the document? Could the Founding Fathers have inadvertently left open a legal door to fascism?

It should be remembered that while Gödel was supremely logical, he was also supremely paranoid and not a little naive. There was something sweetly Pnin-like about him. He believed in ghosts; had a morbid dread of refrigerator gases; pronounced the pink flamingo that his hoydenish wife placed outside his window *furchtbar herzig* (awfully charming); and was convinced, based on nose measurements he had made on a newspaper photograph, that General MacArthur had been replaced by an impostor. His paranoia, though, was decidedly tragic. "Certain forces" were at work in the world "directly submerging the good," he believed.

So was the contradiction in the Constitution, or was it in Gödel's head? I decided to consult a distinguished professor of constitutional law, Laurence Tribe. Besides teaching at Harvard Law School, Tribe had a knack for algebraic topology in his undergraduate days.

"It's unlikely that Gödel could have found anything of the form p and not-p in the Constitution," Tribe told me. "What might have bothered him, though, was Article V, which places almost no substantive constraints on how the Constitution can be amended. He could have interpreted this to mean that as long as an amendment is proposed and approved in the prescribed way, it automatically becomes part of the Constitution, even if it would eliminate the essential features of a republican form of government and obliterate virtually all the protections of human rights."

Tribe continued, "But if I'm correct, Gödel's concern rested on something of a non sequitur. The idea that any constitution could so firmly entrench a set of basic rights and principles as to make them invulnerable to orderly repudiation is unrealistic. Nations like India that purport to make certain basic principles unamendable have in no sense experienced greater fidelity to human rights or democracy than has the United States."

A couple of other legal scholars I spoke to concurred with Tribe that Article V must have been what was vexing Gödel. But the mystery of whether he found something genuinely kinky in the Constitution remains a bit like the mystery of whether Fermat really had a "marvelous proof" of his last theorem. How I wish I had been the judge at that citizenship hearing. Imagine being presented with the opportunity to lean forward, look this agitated genius in the eye, and say, "Surely you must be joking, Mr. Gödel."

The Law of Least Action

Suppose you are standing on the beach, at some distance from the water. You hear cries of distress. Looking to your left, you see someone drowning. You decide to rescue this person. Taking advantage of your ability to move faster on land than in water, you run to a point at the edge of the surf close to the drowning person, and from there you swim directly toward him. Your path is the quickest one to the swimmer, but it is not a straight line. Instead, it consists of two straight-line segments, with an angle between them at the point where you enter the water.

Now consider a beam of light. Like you, it moves faster through air than through water. If it starts from point A in the air and ends at point B in the water, it will not travel in a straight line. Rather, it will take a straight path from point A to the edge of the water, turn a bit, and then follow another straight path to point B in the water. (This is called refraction.) Just as you did when rescuing the drowning person, the light beam considers its destination, then chooses the trajectory that gets it there in the least time, given its differential rate of progress in the two elements through which it must travel.

But this can't be right, can it? Our explanation for the route taken by the light beam—first formulated by Pierre de Fermat in the seventeenth century as the "principle of least time"—assumes that the light somehow knows where it is going in advance and that it acts purposefully in getting there. This is what's called a teleological explanation.

The idea that things in nature behave in goal-directed ways goes back to Aristotle. A final cause, in Aristotle's physics, is the end, or telos, toward which a thing undergoing change is aiming. To explain a change by its final cause is to explain it in terms of the result it achieves. An efficient cause, by contrast, is that which initiates the process of change. To explain a change by its efficient cause is to explain it in terms of prior conditions.

One view of scientific progress is that it consists in replacing teleological (final cause) explanations with mechanistic (efficient cause) explanations. The Darwinian revolution, for instance, can be seen in this way: traits that seemed to have been purposefully designed, like the giraffe's long neck, were re-explained as the outcome of a blind process of chance variation and natural selection.

Actually, pretty much the opposite has happened in physics. In 1744, the French mathematician and astronomer Pierre-Louis Moreau de Maupertuis put forward a grand teleological principle called the "law of least action," which was inspired by (and possibly stolen from) Leibniz. A more abstract version of Fermat's least-time principle, Maupertuis's law said, in essence, that nature always achieves its ends in the most economical way. And what was this "action" that nature supposedly economizes on? As characterized by Maupertuis, it was a mathematical amalgam of mass, velocity, and distance.

In its original form, the law of least action was too vague to be of much use to science. But it was soon sharpened by the great eighteenth-century mathematician Joseph Lagrange. In 1788, a century after Newton's *Principia*, Lagrange published his celebrated *Mécanique analytique*, which expressed the Newtonian system in terms of the law of least action. In the next century, the Irishman William Rowan Hamilton cast the same final-cause idea into a form—which became known as Hamilton's principle—from which all of Newtonian mechanics and optics could be deduced.

Since then, the law of least action has, in its various guises, continued to be extraordinarily powerful. Einstein's equations of relativity, which replaced Newtonian gravitation, can be derived from an action principle not unlike the one Maupertuis set forth. "The highest and most coveted aim of physical science is to condense all natural phenomena which have been observed and are still to be observed into one simple principle," observed Max Planck, the founder of quantum theory. "Amid the more or less general laws which mark the achievements of physical science during the course of the last centuries, the principle of least action is perhaps that which . . . may claim to come nearest to this ideal final aim of theoretical research."

If the law of least action (or a modern version of it) really does stand at the pinnacle of science, what does this say about the world? Does it mean that there is a purposeful intelligence guiding all things with a minimal expenditure of effort, as Maupertuis, Lagrange, and Hamilton believed?

We have one set of equations that explains the world in terms of efficient causes. We have another set that explains it in terms of final causes. The second set may be simpler than the first and more fruitful in leading to new discoveries. But the two describe the *same* state of affairs and yield the *same* predictions. Therefore, as Planck said, "on this occasion everyone has to decide for himself which point of view he thinks is the basic one." You can be a teleologist if you wish. You can be a mechanist if that better suits your fancy. Or you may be left wondering whether this is yet another metaphysical distinction that does not make a difference.

Emmy Noether's Beautiful Theorem

Suppose we want to say that a certain theory of the world is objectively true. What might this mean? Well, among other things, it means that the theory must be true for all observers, regardless of their point of view. That is, its validity should not depend on where you happen to be standing, or which way you happen to be looking, or what time it is.

A theory that is independent of perspective is said to possess symmetry. In normal parlance, "symmetric" is used to describe objects, not theories. Human faces, snowflakes, and crystals are symmetric in certain ways. A sphere has a greater degree of symmetry than any of these because, regardless of how you rotate it, it retains the same form.

Therein lies a clue as to how symmetry might be defined more abstractly. A thing possesses a symmetry if there is something you can do to it that leaves it looking exactly as it did before. This was the definition that the physicist Hermann Weyl (1885–1955) came up with. A *theory* is said to possess a symmetry if there is something you can do to it—like shifting its coordinates in space or time—that leaves the equations of the theory looking exactly as they did before. A shift in coordinates is like a change in perspective. (By shifting the theory's time coordinate, for example, I shift the perspective from the present into the past or future.) Thus, the more symmetries a theory has, the more universally valid its equations are.

I have just set the stage for one of the most underappreciated discoveries of the last century: For each symmetry that a theory possesses, there is a corresponding law of conservation that must hold in the world it describes. A conservation law is one that states that some quantity can be neither created nor destroyed. If a particular theory is symmetric under *space* translation—that is, if its equations remain unchanged when the spatial perspective is shifted—that implies the law of conservation of *momentum*. Similarly, if the theory is symmetric under *time* translation, that implies the law of conservation of *energy*. Symmetry under shifts in *orientation* implies the law of conservation of *angular momentum*. Other, more subtle symmetries imply still more subtle conservation laws.

The intimate connection between symmetry and conservation is "a most profound and beautiful thing," in the words of Richard Feynman, one that "most physicists still find somewhat staggering." Laws that were once thought to be brute facts about the natural world—like the first law of thermodynamics, which says that energy can be neither created nor destroyed—turn out to be, in effect, preconditions of the possibility of objective knowledge. Whenever we formulate a theory of the world that purports to be valid not just from our own point of view but across an entire range of perspectives, we are implicitly committing ourselves to a law of conservation. This revelation has a decidedly Kantian flavor. But Kant's transcendental reasoning was fuzzy and often fallacious. The symmetry-conservation link, by contrast, was proved with airtight logical rigor, by a woman named Emmy Noether.

Emmy Noether was among the greatest pure mathematicians of the twentieth century. Born in Bavaria in 1882, she obtained a Ph.D. at Göttingen in 1907. Though the equal of such illustrious colleagues as David Hilbert, Felix Klein, and Hermann Minkowski, Noether was, as a woman, barred from holding a full professorship, but she was allowed to give unpaid lectures as a *Privatdozent*. When the Nazis came to power in 1933, Noether, a Jew, was stripped of her semiofficial position at Göttingen. She fled to the United States, where she taught at Bryn Mawr and gave lectures at the Institute for Advanced Study in Princeton. In 1935, she died suddenly from an infection after an operation.

Loud of voice and stout of figure, Noether struck her friend Hermann Weyl as looking like "an energetic, nearsighted washerwoman." In addition to being one of the pioneers of abstract algebra, she had considerable literary gifts, writing poetry, a novel, and an autobiography, and co-authoring a play. Her discovery that symmetries in a theory imply conservation laws was published in 1918. It is sometimes called Noether's theorem.

Does Noether's theorem mean that laws of conservation are not "out there" in the world but merely artifacts of our epistemic practice? Such an idealistic interpretation should be resisted. The world does exert some control over just how symmetric—that is, universal—a true theory of it can be. Some symmetries have failed experimentally. In 1957, for example, Tsung-Dao Lee and Chen Ning Yang were awarded the Nobel Prize in Physics for showing that a certain particle-decay process violated "conservation of parity"—which means that the laws of physics would be slightly different in a universe that was the mirror image of our own.

If the law of conservation of energy were ever shown to fail, the consequences would be more serious. The true theory of the world would then, we know from Noether's theorem, depend on what time it was—which would be a great blow to its "objectivity."

The curious thing is that at many points in the history of science it looked as though the energy-conservation principle *had* failed. Each time, however, it was salvaged by making the concept of energy more general and abstract. What began as a purely mechanical notion eventually came to encompass thermal, electric, magnetic, acoustic, and optical varieties of energy—all, fortunately, interconvertible. With Einstein's theory of relativity, even matter came to be viewed as "frozen" energy.

Sooner than give up energy conservation, Henri Poincaré once observed, we would invent new forms of energy to save it. Thanks to Emma Noether's profoundly beautiful discovery, we know why: the timelessness of physical truth hangs on it.

Is Logic Coercive?

The English cleric and satirist Sydney Smith once observed two women quarreling with each other from their respective attic windows across a narrow street in Edinburgh. "Those two women will never agree," he remarked. "They are arguing from different premises."

That was in the early nineteenth century. Today, the situation is worse. Even sharing the same premises with your interlocutor is no guarantee that you will eventually come to the same conclusion. You must also share the same logic.

The purpose of logic is to distinguish valid forms of argument from faulty ones, called fallacies. If you conduct me from premises that I accept to a conclusion that I dislike by means of a fallacious argument, I am under no obligation to embrace that conclusion. If I, on the other hand, take you from premises you accept to a conclusion you dislike by means of a logically valid argument, then you are compelled to embrace the conclusion.

"Compelled by what?" you ask. By the court of rationality, I reply. The fact that my argument is logically valid means that the premises cannot be true without the conclusion being true; so if you believe the premises, you *have* to believe the conclusion, or you are being irrational. "What's so bad about being irrational?" you ask defiantly. I then dig myself in deeper by giving you reasons for accepting my reasons, which leave you similarly unmoved. What I'd like to do, though, is turn my logic into a club and use it to beat you into submission.

The impotence of logic was celebrated by Robert Nozick in his 1981 book, *Philosophical Explanations*. "Why are philosophers intent on forcing others to believe things?" he asked. "Is that a nice way to behave toward someone?" What logicians really wish they had, Nozick darkly suggests, is an argument that sets up reverberations in the brain so that if the person refuses to accept the conclusion, he *dies*.

One basic logical principle is the law of noncontradiction, which says that a proposition and its negation cannot both be true. In fact, there are many people who violate this law without realizing it. They believe propositions p, q, r, and s while unaccountably failing to notice that q, r, and s jointly entail not-p. Should you bring this to their attention, they might eject proposition p from their creed. More likely, they will try to dodge the charge of inconsistency by quibbling about meanings. ("What I said was that neutral countries should not be *invaded*. That was an *incursion*, not an *invasion*!")

But what if your interlocutor invokes the spirit of Walt Whitman and says, "Do I contradict myself? Very well then I contradict myself." (The physicist Niels Bohr once came close to this. A colleague, seeing a horseshoe over Bohr's office door, said, "You don't really believe in that stuff, do you?" Bohr replied, "No, but I hear it works even for those who don't believe.") What do you say to such a person?

"This would vitiate all science" is what W. V. Quine, one of the preeminent logicians of the twentieth century, would say: "Any conjunction of the form p and not-p logically implies every sentence whatever; therefore acceptance of one sentence and its negation as true would commit us to accepting every sentence as true, and thus forfeiting all distinction between true and false." To see what Quine means, suppose you believe both p and not-p. Because you believe p, you must also believe p or q, where q is any arbitrary proposition. But from p or q and not-p, it obviously follows that q. Hence *any* arbitrary proposition is true. (The Latin phrase for this is *ex contradictione quodlibet*.)

The idea that a contradiction is bad because absolutely anything follows from it might seem strange to a non-logician. Bertrand Russell was once trying to get this very point across at a public lecture when a heckler interrupted him. "So prove to me that if two plus two is five,

I'm the Pope," the heckler said. "Very well," Russell replied. "From 'two plus two equals five' it follows, subtracting three from each side, that two equals one. You and the Pope are two, therefore you are one."

Formal logic, as philosophy departments have traditionally taught it, seems "coercive" to some. The philosopher Ruth Ginzberg has taken issue with the law of *modus ponens*, which sanctions inferences of the form "if p, then q; but p; therefore q." Ginzberg maintains that *modus ponens* is used by males to marginalize women—who are supposedly less likely to recognize its supposed validity—as "irrational." In a different vein, some would-be reformers have suggested that fallacies be thought of as "failures of cooperation" rather than as errors of reasoning. Others advocate the adoption of "principles of charity": if, for example, your interlocutor's argument is a muddle, try to reconstrue it in a way that makes it valid.

Whether this approach is a better route to truth is arguable. But the new niceness in logic certainly threatens to take some of the zest out of polemics. Coercion by logic—or by a simulacrum of logic—can be a marvelous blood sport. One thinks, for example, of the great confrontation between Diderot and the Swiss mathematician Leonhard Euler before the court of Catherine the Great in 1773. Diderot, an atheist, was almost entirely ignorant of mathematics. Euler, a devout Christian, approached the *philosophe*, bowed, and said very solemnly, "Sir, $(a + b^n) / n = x$, hence God exists. Reply!" Delighted laughter broke out on all sides as Diderot crumpled before this stunning inference.

The next day Diderot asked Catherine for permission to return to France, a request to which the empress graciously consented.

Newcomb's Problem and the Paradox of Choice

The philosopher Robert Nozick (1938–2002) was famous as the author of *Anarchy, State, and Utopia*. This closely reasoned defense of the minimal state, published in 1974, resonated with libertarian types everywhere and became something of a bible for many Warsaw bloc dissidents. But Nozick never thought of himself as a political philosopher. (*Anarchy, State, and Utopia* was an "accident," he claimed; he was prodded to write it by the appearance of his Harvard colleague John Rawls's book *A Theory of Justice*.) He was more interested in rational choice and free will. What had launched his philosophical career was a wonderful paradox that involved both these topics. Nozick did not invent this paradox himself. It was thought up by a California physicist named William Newcomb and reached Nozick by way of a mutual friend, the Princeton mathematician David Kruskal, at a cocktail party ("the most consequential party I have attended," said Nozick, who was by no means party shy).

Nozick wrote about this paradox in his dissertation and then published an article on it in 1969 titled "Newcomb's Problem and Two Principles of Choice." The result was, as *The Journal of Philosophy* put it, "Newcombmania." Suddenly everyone in the philosophical world was writing and arguing about Newcomb's problem. Now, half a century later, and despite the best efforts of Nozick and dozens of other philosophers, the paradox remains just as perplexing—and divisive—as it was when it was first conceived.

Newcomb's problem goes like this. There are two closed boxes on the table, box A and box B. Box A contains a thousand dollars. Box B contains either a million dollars or no money at all. You have a choice between two actions: (1) taking what is in both boxes; or (2) taking just what is in box B.

Now here comes the interesting part. Imagine a Being that can predict your choices with high accuracy. You can think of this Being as a genie, or a superior intelligence from another planet, or a supercomputer that can scan your mind, or a very shrewd psychologist, or God. The Being has correctly predicted your choices in the past, and you have enormous confidence in his predictive powers. Yesterday, the Being made a prediction as to which choice you are about to make, and it is this prediction that determines the contents of box B. If the Being predicted that you will take what is in both boxes—action 1—he put nothing in box B. If he predicted that you will take only what is in box B—action 2—he put a million dollars in box B. You know these facts, he knows you know them, and so on.

So, do you take both boxes, or only box B?

Well, obviously you should take only box B, right? For if this is your choice, the Being has almost certainly predicted it and put a million dollars in box B. If you were to take both boxes, the Being would almost certainly have anticipated this and left box B empty. Therefore, with very high likelihood, you would get only the thousand dollars in box A. The wisdom of the one-box choice seems confirmed when you notice that of all your friends who have played this game, the one-boxers among them are overwhelmingly millionaires, and the two-boxers are overwhelmingly not.

But wait a minute. The Being made his prediction yesterday. Either he put a million dollars in box B, or he didn't. If it's there, it's not going to vanish just because you choose to take both boxes; if it's not there, it's not going to materialize suddenly just because you choose only box B. Whatever the Being's prediction, you are guaranteed to end up a thousand dollars richer if you choose both boxes. Choosing just box B is like leaving a thousand-dollar bill lying on the sidewalk. To make the logic of the two-box choice even more vivid, suppose the backs of the boxes are made of glass and your wife is sitting on the other side of the table. She can plainly see what's in each box. You know which choice she wants you to make: Take both boxes!

So you can see what's paradoxical about Newcomb's problem. There are two powerful arguments as to what choice you should make—arguments that lead to precisely opposite conclusions. The first argument, the one that says you should take just box B, is based on the principle of *maximizing expected utility*. If the Being is, say, 99 percent accurate in his predictions, then the expected utility of taking both boxes is $0.99 \times \$1,000 + 0.01 \times 1,001,000 = \$11,000$. The expected utility of taking only box B is $0.99 \times \$1,000,000 + 0.01 \times \$0 = \$990,000$. The two-box argument is based on the principle of *dominance*, which says that if one action leads to a better outcome than another in every possible state of affairs, then that's the action to take. These principles can't both be right, on pain of contradiction. And they play the devil with intuition.

"I have put this problem to a large number of people, both friends and students in class," Nozick wrote in the 1969 article. "To almost everyone it is perfectly clear and obvious what should be done. The difficulty is that these people seem to divide almost evenly on the problem, with large numbers thinking that the opposing half is just being silly." When Martin Gardner presented Newcomb's problem in 1973 in his *Scientific American* column, the enormous volume of mail it elicited ran in favor of the one-box solution by a five-to-two ratio. (Among the correspondents was Isaac Asimov, who perversely plumped for the two-box choice as an assertion of his free will and a snub to the predictor, whom he identified with God.)

Newcomb, the begetter, was a one-boxer. Nozick himself started out as a lukewarm two-boxer, despite being urged by the decision theorists Maya Bar-Hillel and Avishai Margalit to "join the millionaires' club" of one-boxers. By the 1990s, however, Nozick had arrived at the unhelpful view that both arguments should be given some weight in deciding which action to take. After all, he reasoned, even resolute two-boxers will become one-boxers if the amount in box A is reduced to $1, and all but the most die-hard one-boxers will become two-boxers if it is raised to $900,000, so nobody is completely confident in either argument.

Other philosophers refuse to commit themselves to one choice or the other on the grounds that the whole setup of Newcomb's problem is nonsensical: If you really have free will, they argue, then how could any Being accurately predict how you would choose between two equally rational actions, especially when you know that your choice has been predicted before you make it?

Actually, though, the predictor does not have to be all that accurate to make the paradox work. We've already seen that the readers of *Scientific American* favored the one-box solution by a five-to-two ratio. So, for that crowd at least, a perfectly ordinary Being could achieve better than 70 percent accuracy by always predicting that the one-box choice would be made. A psychologist might improve that accuracy rate by keeping track of how women, left-handers, Ph.D.'s, Republicans, and so on tended to choose. If I were playing the game with a human predictor whose high accuracy depended on such statistics, I should certainly opt for the contents of both boxes. On the other hand, if the Being were supernatural—a genie or God or a genuine clairvoyant—I would probably take only box B, out of concern that my choice might affect the Being's prediction through some sort of backward causation or timeless omniscience. I would also have to wonder whether I was really choosing freely.

The quantity and ingenuity of the resolutions proposed for Newcomb's problem over the years have been staggering. (It has been linked to Schrödinger's cat in quantum mechanics and Maxwell's demon in thermodynamics; more obviously, it is analogous to the prisoner's dilemma, where the other prisoner is your identical twin who will almost certainly make the same choice you do to cooperate or defect; and— more terrifyingly to some—it is at the core of Roko's basilisk, a reputedly self-realizing thought experiment in which the stakes include eternal torment by a godlike future AI being.) Yet none of these resolutions have been completely convincing, so the debate goes on. Could Newcomb's problem turn out to have the longevity of Zeno's paradoxes? Will philosophers still be vexing over it twenty-five hundred years from now, long after *Anarchy, State, and Utopia* is forgotten? If so, it is sad that Nozick, the man who put Newcomb's problem on the intellectual map, should not be the one to enjoy eponymous immortality. "It is a beautiful problem," he wrote, in a melancholy vein. "I wish it were mine."

The Right Not to Exist

At the turn of the millennium, France's highest court handed down a ruling of great moral and even metaphysical interest. The court declared that a seventeen-year-old boy was entitled to compensation for being born. Because he contracted German measles from his mother while she was pregnant with him—both a doctor and a laboratory failed to diagnose her illness—the boy grew up deaf, mentally retarded, and nearly blind.

From the parents' point of view, the court's decision made eminent sense. Had they known about the measles and the associated risks to the fetus, they could have aborted, waited a few months, and then started another pregnancy that would have issued in a healthy baby. From the perspective of the seventeen-year-old boy, however, the logic of the decision might have seemed a little peculiar. The correct diagnosis, after all, would have led to an outcome in which he did not exist. Would this really have been better for him?

The very idea of judging one's life better or worse than complete nonexistence strikes some philosophers as absurd. Bernard Williams, for example, argued that a person simply "cannot think egoistically of what it would be for him never to have existed." Others, like the late Derek Parfit, contend that it at least makes sense to say of a life that it is worth living or not worth living. If the former, that life is better than nothing; if the latter, it is worse than nothing. Yet even Parfit shrank from the implication that a person whose life is not worth living would have been better off if he or she had never existed.

Some people feel that any kind of life, no matter how awful, is still better than nothing. In his essay "Death," Thomas Nagel characterized this view: "There are elements which, if added to one's experience, make life better; there are other elements which, if added to one's experience, make life worse. But what remains when these are set aside is not merely neutral; it is emphatically positive. Therefore life is worth living even when the bad elements of experience are plentiful, and the good ones too meager to outweigh the bad ones on their own."

If every life is worth living, then it can never be wrong to bring a child into the world, no matter how defective that child might be. It is often observed that even a child with Down syndrome can live a happy life. But there are other genetic conditions that have far worse consequences. Boys afflicted with Lesch-Nyhan syndrome, for instance, not only suffer mental retardation and excruciating physical pain; they also compulsively mutilate themselves. Most of us feel that it would be wrong knowingly to conceive such a child; in fact, that one has a duty not to do so.

But there is a curious asymmetry here. Consider a couple who know that if they decide to have a child, it will likely have a happy life. Does this couple have a duty to go ahead and conceive a child? Most of us would say no. But why? After all, if the misery of a possible child creates an ethical obligation not to bring it into the world, shouldn't the happiness of a possible child create an ethical obligation to bring it into the world? Why should the well-being of a possible child enter our moral calculus in one case but not the other?

Moral philosophers have yet to come up with a satisfying explanation for this asymmetry. Witness the rather tortuous attempt of Peter Singer: "Perhaps the best one can say—and it is not very good—is that there is nothing directly wrong in conceiving a child who will be miserable, but once such a child exists, since its life can contain nothing but misery, we should reduce the amount of pain in the world by an act of euthanasia. But euthanasia is a more harrowing process for the parents and others involved than non-conception. Hence we have an indirect reason for not conceiving a child bound to have a miserable existence." For Singer, it is not the life prospects of the possible child— whether happy or miserable—that create ethical obligations for its potential parents to conceive or not to conceive that child. What justifies treating these cases asymmetrically, in his view, is the unhappiness that would be experienced by the actual parents over the need to euthanize the miserable child, should they choose to conceive it.

If the French boy's life is even marginally worth living, then it was fortunate for him that his mother's case of German measles was not diagnosed. Let's suppose, however, that the boy's life is not worth living. One might say that the doctor who failed to detect the measles infringed on the boy's right not to have a miserable life. Yet this is a right that, in his case, could not possibly have been fulfilled: owing to contingent features of the human reproductive system, a child conceived a few months later would have had a different genetic identity and hence a different personal identity.

As for the child who might have been conceived by the French couple several months later had the measles been diagnosed, there are two ways of looking at his or her plight. If you believe only in the actual world, then the closest he or she came to reality was as a pair of unconnected (and now long-defunct) gametes. Nothing can be fortunate or unfortunate for such an entity. If, on the other hand, you believe in the reality of possible worlds—as the influential philosopher David Lewis avowedly did—then there are many, many possible worlds that contain versions of this child, and within each of those worlds the respective child is fortunate to exist.

Whether or not you agree with Thomas Nagel's cheerful assertion that "all of us . . . are fortunate to have been born," he certainly got it right when he added, "It cannot be said that not to be born is a misfortune." And when the chorus in *Oedipus at Colonus* gloomily declares, "Not to be born is best of all," the appropriate riposte is, how many are so lucky?

Can't Anyone Get Heisenberg Right?

In the *Routledge Encyclopedia of Philosophy*, "Heisenberg, Werner" lies between "Heidegger, Martin" and "Hell." That is precisely where he belongs. Heisenberg, one of the inventors of quantum mechanics, was the leader of Hitler's atomic bomb project during World War II. After the war, he claimed that he had deliberately sabotaged the Nazi bomb effort. Many believed him. But it seems more likely that his failure was due not to covert heroism but to incompetence.

Heisenberg (1901–1976) was a wonderful physicist. At the age of twenty-four, in a rapture on a rock overlooking the North Sea, he had an insight that revolutionized our understanding of the subatomic world. Two years later he announced, in what is probably the most quoted paper in the history of physics, his "uncertainty principle." Yet his reasoning was far from transparent. Even the greatest physicists admit to bafflement at his mathematical non sequiturs and leaps of logic. "I have tried several times to read [one of his early papers]," confesses the Nobel laureate Steven Weinberg, "and although I think I understand quantum mechanics, I have never understood Heisenberg's motivations for the mathematical steps."

Though he might have been a magician as a theorist, Heisenberg was something of a dunce at applied physics. His doctoral exam in 1923 was a disaster. Asked about it many years later by Thomas Kuhn, he gave the following account (his examiner was the experimental physicist Wilhelm Wien): "Wien asked me . . . about the Fabry-Perot interferometer's resolving power . . . and I'd never studied that . . . Then he got annoyed and asked about a microscope's resolving power. I didn't know that. He asked me about a telescope's resolving power, and I didn't know that either . . . So he asked me how a lead storage battery operates and I didn't know that . . . I am not sure whether he wanted to fail me." When, during the war, Heisenberg tried to determine how much fissionable uranium would be necessary for a bomb, he botched the calculation and came up with the impossible figure of several tons. (The Hiroshima bomb required only fifty-six kilograms.) This is not the kind of scientist you want to put in charge of a weapons project.

Those who wish to stress the supposed murkiness of Heisenberg's wartime motives often reach for a metaphor from his physics: the uncertainty principle. Michael Frayn did it in *Copenhagen*, his play about a mysterious 1941 encounter between Heisenberg and Bohr. Thomas Powers did it in *Heisenberg's War*, the 1993 book that defended Heisenberg's claim to have destroyed the Nazi bomb project from within. David C. Cassidy did it in the very title of his 1991 biography of Heisenberg, *Uncertainty*. They should all have known better.

And they're hardly alone. No scientific idea from the last century is more fetishized, abused, and misunderstood—by the vulgar and the learned alike—than Heisenberg's uncertainty principle. The principle doesn't say anything about how precisely any particular thing can be known. It does say that some *pairs* of properties are linked in such a way that they cannot both be measured precisely at the same time. In physics, these pairs are called canonically conjugate variables. One such pair is position and momentum: the more precisely you locate the position of a particle, the less you know about its momentum (and vice versa). Another is time and energy: the more precisely you know the time span in which something occurred, the less you know about the energy involved (and vice versa).

How could this principle of physics be applied to Heisenberg the man? In the postscript to his play *Copenhagen*, Frayn writes, "There is not one single thought or intention of any sort that can ever be precisely established." Well, maybe, but the uncertainty principle applies to *pairs* of properties. In Heisenberg's case, the relevant pair is motivation and competence. How willing was he to help Hitler? How competent was he to produce an atomic bomb? But notice that there is a positive relationship between our knowledge of one and of the other: the more certain we become that Heisenberg was willing to serve the Third Reich, the more certain we become that he was incompetent to produce a bomb. This is not the uncertainty principle but its exact opposite. Evidently, knavishness and incompetence are not canonically conjugate variables.

A more banal misuse of Heisenberg's principle can be found in the social sciences. There the principle is often taken to mean that the very act of observing a phenomenon inevitably alters that phenomenon in some way; that is why, say, Margaret Mead could never know the sexual mores of the Samoans—her very presence on the island distorted what she was there to observe. Postmodern theorists like Stanley Aronowitz have invoked the uncertainty principle as proof of the unstable hermeneutics of subject-object relations, arguing that it undermines science's claim to objectivity.

Even physicists show considerable uncertainty about what the uncertainty principle really means. Dozens of different interpretations have been proposed over the years. Some locate the uncertainty in an inherent and ineliminable clumsiness in the act of measurement itself. How do you learn the position of an electron with great accuracy? By bouncing a photon off it. But because the electron is quite tiny, the photon must have a comparably tiny wavelength and thus a very great energy (since wavelength and energy are inversely related). So, the photon will impart a random "kick" to the electron that will affect its momentum in an unknowable way.

Heisenberg himself opted for this kind of interpretation, which is called epistemic, because it places the burden of uncertainty on the knower. Niels Bohr, by contrast, plumped for an "ontic" interpretation, attributing the uncertainty not to the knower and his measurement apparatus but to reality itself. Familiar concepts like position and momentum simply do not apply at the quantum level, Bohr argued. The contemporary physicist Roger Penrose has declared himself unhappy with the whole gamut of interpretations of Heisenberg's principle while admitting he has nothing better to replace them with just now.

From a mathematical point of view, there is nothing the least bit problematic about Heisenberg's uncertainty principle. If you try to translate the sentence "Electron e is exactly at position x with a momentum of exactly p" into the formal language of quantum theory, you get ungrammatical gibberish, just as you would if you tried to translate "the round square" into the language of geometry. It is only when you try to make sense of the principle philosophically that the waters begin to rise up around you.

Some decades ago, the Princeton physicist John Archibald Wheeler began to wonder whether Heisenberg's uncertainty principle might not have some deep connection to Gödel's incompleteness theorem (probably the second-most-misunderstood discovery of the twentieth century). Both, after all, seem to place inherent limits on what it is possible to know. But such speculation can be dangerous. "Well, one day,"

Wheeler recounts, "I was at the Institute for Advanced Study, and I went to Gödel's office, and there was Gödel. It was winter and Gödel had an electric heater and had his legs wrapped in a blanket. I said, 'Professor Gödel, what connection do you see between your incompleteness theorem and Heisenberg's uncertainty principle?' And Gödel got angry and threw me out of his office."

Overconfidence and the Monty Hall Problem

Can you spot a liar? Most people think they are rather good at this, but they are mistaken. In study after study, subjects asked to distinguish between videotaped liars and truth tellers have performed miserably at the task, scoring little better than chance. That goes even for those who were especially sure of their expertise in catching out lies—police detectives, for instance.

Human beings, as it turns out, have too much faith in themselves. The detection of falsehood is scarcely the only domain where we overestimate our abilities. A survey of British motorists once revealed that 95 percent of them thought they were better-than-average drivers. Similarly, most people think they are likely to live longer than the mean. In a classic 1977 paper published in the *Journal of Experimental Psychology*, Baruch Fischhoff, Paul Slovic, and Sarah Lichtenstein reported that people often pronounce themselves absolutely certain of beliefs that are untrue. Subjects would declare themselves 100 percent sure that, say, the potato originated in Ireland, when it actually came from Peru.

Overconfidence is nearly universal. But is it distributed equally? Evidently not. In a 1999 paper in the *Journal of Personality and Social Psychology*, David A. Dunning and Justin Kruger drew a poignant conclusion from their research: The most incompetent people have the most inflated notion of their abilities. "Not only do they reach erroneous conclusions and make unfortunate choices," the two psychologists observed, "but their incompetence robs them of the ability to realize it."

Dunning and Kruger administered three sorts of tests to their subjects: logic, English grammar, and humor (where ratings of jokes were judged against those of a panel of professional comedians). In all three, the subjects who did worst were the most likely to "grossly overestimate" how well they had performed. Those who scored in the twelfth percentile in the logic test, for example, imagined that their overall skill in logic was at the sixty-eighth percentile.

Now, if you are among the competent, you might derive some consolation from this research, since it implies that you are unlikely to be grossly overconfident. But perhaps you simply *imagine* that you are among the competent—precisely because you suffer from the overconfidence of the incompetent.

And there is more to worry about. Overconfidence may decrease with competence, but other studies show that it increases with knowledgeability; that is, the more specialized information you have about something, the more likely you are to be overconfident in your judgments about it. Overconfidence also tends to rise with the complexity of the problem. This means that experts reasoning about difficult matters—doctors, engineers, financial analysts, academics, even the pope when he is not speaking ex cathedra—are apt to be seriously overconfident in the validity of their conclusions.

Let me illustrate the point with an anecdote. (Who was it who said that a social scientist is someone who thinks that the plural of "anecdote" is "data"?) Paul Erdős (1913–1996) was one of the supreme mathematicians of the last century. He was also one of the world's leading experts on probability theory; indeed, something he invented called the probabilistic method is often simply referred to as the Erdős method—thus making his name synonymous with probability.

In 1991, Erdős found himself befuddled when the *Parade* magazine columnist Marilyn vos Savant published a probability puzzle called the Monty Hall problem, named after the original emcee of the TV game show *Let's Make a Deal*. It goes like this. There are three doors onstage, labeled A, B, and C. Behind one of them is a sports car; behind the other two are goats. You get to choose one of the doors and keep whatever is behind it. Let's suppose you choose door A. Now, instead of showing you what's behind it, Monty Hall slyly opens door B and reveals . . . a goat. He then offers you the option of switching to door C. Should you take it? (Assume, for the sake of argument, that you are indifferent to the charm of goats.)

Counterintuitively enough, the answer is that you should switch, because a switch increases your chance of winning from one-third to two-thirds. Why? When you initially chose door A, there was a one-third chance that you would win the car. Monty's crafty revelation that there's a goat behind door B furnishes no new information about what's behind the door you already chose—you already *know* one of the other two doors has to conceal a goat—so the likelihood that the car is behind door A remains one-third. Which means that with door B eliminated, there is a two-thirds chance that the car is behind door C.

But Erdős insisted to his friends it wasn't so. His intuition told him that switching should make no difference in the odds. And this peerless authority on probability was confident in his intuition—so confident that he remained in high dudgeon for several days until a mathematician at Bell Labs made him see his error.

One final generalization can be drawn from the psychological literature: high levels of confidence are usually associated with high levels of *over*confidence. The gap between conviction and truth seems to be greatest in the case of those judgments about which one feels most certain. Who knows? The most overconfident judgment in history might turn out to be *cogito, ergo sum*.

The Cruel Law of Eponymy

"Who is buried in Grant's Tomb?" That was the bonus question Groucho Marx used to put to unfortunate contestants on his 1950s quiz show *You Bet Your Life*. It sounds like a giveaway, but beware: questions of this form can be treacherous. Consider: Who discovered Bayes' theorem? Who discovered Giffen's paradox? Who discovered the Pythagorean theorem? Who discovered America? If your answers were, respectively, Bayes, Giffen, Pythagoras, and Amerigo Vespucci, then no box of Snickers for you.

The practice of naming things after people (real or mythical) who are associated with them is called eponymy. There are eponymous

words, like "guillotine," "bowdlerize," and "sadism." There are eponymous place-names, like Pennsylvania and the Peloponnesus. And
there are eponymous compound expressions, like "Copernican system"
and "Halley's comet." When such expressions occur in the sciences,
the presumption is that the thing designated was discovered by the
scientist whose name is affixed to it. That presumption is nearly always false.

If you think I exaggerate, you are obviously not familiar with Stigler's law of eponymy. This law, which in its simplest form states that
"no scientific discovery is named after its original discoverer," was so
dubbed by the historian/statistician Stephen Stigler. An immodest act
of nomenclature? Not really. If Stigler's law is true, its very name implies that Stigler himself did not discover it. By explaining that the
credit belongs instead to the great sociologist of science Robert K.
Merton, Stigler not only wins marks for humility, he makes the law to
which he has lent his name self-confirming.

What explains the truth of Stigler's law? One might start with
Merton's famous hypothesis that "all scientific discoveries are in principle 'multiples.'" Perhaps, for some reason, a discovery invariably gets
named for the wrong one of its multiple discoverers.

But Stigler's law is more interesting than that. Take the Pythagorean theorem. Pythagoras was not one of its discoverers. The theorem
was known before him and proved after him; moreover, he might even
have been unaware of its geometric import. Such radical misnamings
abound. In checking out Stigler's suggestion that Giffen's paradox
("demand for some goods increases with price") was undreamed of by
its eponym, the economist Robert Giffen, I chanced upon an encyclopedia entry for Sir Thomas Gresham, the sixteenth-century Englishman
after whom Gresham's law ("bad money drives out good") is named.
"It was thought that Gresham was the first to state the principle," the
entry reads, "but it has been shown that it was stated long before his
time and that he did not even formulate it."

Such eponymic blunders might be the exception rather than the
rule if historians of science were in charge of labeling scientific discoveries. But they are not; it's practicing scientists who make the decisions, and for all their vaunted rigor most of them have no historical
expertise. As Stigler observes in his book *Statistics on the Table*, "Names

are rarely given and never generally accepted unless the namer . . . is remote in time or place (or both) from the scientist being honored." That is to ensure the appearance of impartiality. After all, having a theorem or a comet named for you confers something like intellectual immortality; such an honor must be perceived by the community of scientists as being based on merit, not on national affiliation or personal friendship or political pressures.

Given that "eponyms are only awarded after long time lags or at great distances, and then only by active (and frequently not historically well informed) scientists with more interest in recognizing general merit than an isolated achievement," Stigler concludes, "it should not then come as a surprise that most eponyms are inaccurately assigned, and it is even possible (as I have boldly claimed) that all widely accepted eponyms are, strictly speaking, wrong."

The great power of Stigler's law of eponymy can be illustrated by applying it to a particular case: the formula for the probability distribution known as the bell curve. Because this is also called the Gaussian distribution, one can infer from Stigler's law that Gauss was not its discoverer. Sure enough, in an 1809 book, Gauss cites Laplace in connection with the distribution, and in fact Laplace did touch on it as early as 1774. But the distribution is also sometimes called the Laplace or Laplace-Gauss distribution, so it can further be inferred from Stigler's law that Laplace was not its discoverer either. Indeed, current scholarship traces its origin to a 1733 publication by Abraham de Moivre.

Oddly, I have found that Stigler's law is valid even for pseudo-eponyms. Take "crap." People often claim that this word eponymously derives from Thomas Crapper, the celebrated Victorian inventor of the flush toilet. But this etymology is spurious: the word "crap" in its excremental sense entered Middle English from Old French. Nevertheless, the mere fact that Crapper is folkishly linked to "crap" suggests, by Stigler's law, that he was not the original inventor of the flush toilet. And, *mirabile dictu*, that turns out to be the case: the flush toilet was designed by Sir John Harington in the court of Elizabeth I.

I could give more examples, but lunchtime approaches, and I am looking forward to eating something that I feel quite certain was not invented by the fourth Earl of Sandwich.

The Mind of a Rock

Most of us have no doubt that our fellow humans are conscious. We are also pretty sure that many animals have consciousness. Some, like the great ape species, even seem to possess self-consciousness, like us. Others, like dogs and cats and pigs, may lack a sense of self, but they certainly appear to experience inner states of pain and pleasure. About smaller creatures, like mosquitoes, we are not so sure; certainly we have few compunctions about killing them. As for plants, they obviously do not have minds, except in fairy tales. Nor do nonliving things like tables and rocks.

All that is common sense. But common sense has not always proved to be such a good guide in understanding the world. And the part of our world that is most recalcitrant to our understanding at the moment is consciousness itself. How could the electrochemical processes in the lump of gray matter that is our brain give rise to—or, even more mysteriously, *be*—the dazzling Technicolor play of consciousness, with its transports of joy, its stabs of anguish, and its stretches of mild contentment alternating with boredom? This has been called "the most important problem in the biological sciences" and even "the last frontier of science." It engrosses the intellectual energies of a worldwide community of brain scientists, psychologists, philosophers, physicists, computer scientists, and even, from time to time, the Dalai Lama.

So vexing has the problem of consciousness proved that some of these thinkers have been driven to a hypothesis that sounds desperate, if not downright crazy. Perhaps, they say, mind is not limited to the brains of some animals. Perhaps it is ubiquitous, present in every bit of matter, all the way up to galaxies, all the way down to electrons and neutrinos, not excluding medium-sized things like a glass of water or a potted plant. Moreover, it did not suddenly arise when some physical particles on a certain planet chanced to come into the right configuration; rather, there has been consciousness in the cosmos from the very beginning of time.

The doctrine that the stuff of the world is fundamentally mind stuff goes by the name of panpsychism. Several decades ago, the American philosopher Thomas Nagel showed that it is an inescapable consequence of some quite reasonable premises. First, our brains consist of material particles. Second, these particles, in certain arrangements, produce subjective thoughts and feelings. Third, physical properties alone cannot account for subjectivity. (How could the ineffable experience of tasting a strawberry ever arise from the equations of physics?) Now, Nagel reasoned, the properties of a complex system like the brain don't just pop into existence from nowhere; they must derive from the properties of that system's ultimate constituents. Those ultimate constituents must therefore have subjective features themselves—features that, in the right combinations, add up to our inner thoughts and feelings. But the electrons, protons, and neutrons making up our brains are no different from those making up the rest of the world. So the entire universe must consist of little bits of consciousness.

It is sometimes argued that consciousness is an "emergent" property—that it arises from the interactions of neurons in our brains the way, say, liquidity arises from the interactions of nonliquid molecules. But the analogy is faulty. Facts about liquidity, however difficult they are to predict, are still logically dependent on physical facts about individual molecules. Consciousness is not like this. Its subjective features can't be derived from lower-level physical facts. The emergence of mind stuff from physical stuff would be a kind of "brute" emergence that is nowhere to be found in science. So the panpsychists insist.

Nagel himself stopped short of embracing panpsychism, but today it is enjoying something of a vogue. The Australian philosopher David Chalmers, the British philosopher Galen Strawson, and the Oxford physicist Roger Penrose have all spoken on its behalf. Others, like the American philosopher John Searle, find the very notion absurd.

The skeptics of panpsychism have a variety of misgivings. How, they ask, could bits of mind dust, with their presumably simple mental states, combine to form the kinds of complicated experiences we humans have? After all, when you put a bunch of people in the same room, their individual minds do not form a single collective mind. (Or do they?) Then there is the inconvenient fact that you can't scientifically test the claim that, say, the moon is having mental experiences. (But the same applies to people: How could you prove that your fellow office workers aren't unconscious robots, like Commander Data on *Star Trek*?) Finally, there is the sheer loopiness of the idea that something like a photon could have proto-emotions, proto-beliefs, and proto-desires. What could the content of a photon's desire possibly be? "Perhaps it wishes it were a quark," one anti-panpsychist cracked.

Panpsychism may be easier to parody than to refute. But even if it proves a cul-de-sac in the quest to understand consciousness, it might still help rouse us from a certain parochiality in our cosmic outlook. We are biological beings. We exist because of self-replicating chemicals. We detect and act on information from our environment so that the self-replication will continue. As a by-product, we have developed brains that, we fondly believe, are the most intricate things in the universe. We look down our noses at brute matter.

Take that rock over there. It doesn't seem to be doing much of anything, at least to our gross perception. But at the microlevel it consists of an unimaginable number of atoms connected by springy chemical bonds, all jiggling around at a rate that even our fastest supercomputer might envy. And they are not jiggling at random. The rock's innards "see" the entire universe by means of the gravitational and electromagnetic signals it is continuously receiving. Such a system can be viewed as an all-purpose information processor, one whose inner dynamics mirror any sequence of mental states that our brains might run through. And where there is information, says the panpsychist, there is consciousness. In David Chalmers's slogan, "Experience is information from the inside; physics is information from the outside."

Of course, the rock doesn't exert itself as a result of all this "thinking." Why should it? Its existence, unlike ours, doesn't depend on the struggle to survive and self-replicate. It is indifferent to the prospect of being pulverized. If you are poetically inclined, you might think of the rock as a purely contemplative being. And you might draw the moral that the universe is, and always has been, saturated with mind, even though we snobbish Darwinian-replicating latecomers are too blinkered to notice.

PART IX

God, Sainthood, Truth, and Bullshit

21

Dawkins and the Deity

Richard Dawkins, erstwhile holder of the interesting title of Simonyi Professor for the Public Understanding of Science at Oxford University, is a master of scientific exposition and synthesis. When it comes to his own specialty, evolutionary biology, there is none better. But the purpose of his bestselling 2006 book, *The God Delusion*, is not to explain science. It is rather, as he tells us, "to raise consciousness," which is quite another thing.

The nub of Dawkins's consciousness-raising message is that to be an atheist is a "brave and splendid" aspiration. Belief in God is not only a delusion, he argues, but a "pernicious" one. On a scale of 1 to 7, where 1 is certitude that God exists and 7 is certitude that God does not exist, Dawkins rates himself a 6: "I cannot know for certain but I think God is very improbable, and I live my life on the assumption that he is not there."

Dawkins's case against religion follows an outline that goes back to Bertrand Russell's classic 1927 essay "Why I Am Not a Christian." First, discredit the traditional reasons for supposing that God exists. ("God" is here taken to denote the Judeo-Christian deity, presumed to be eternal, all-powerful, all good, and the creator of the world.) Second, produce an argument or two supporting the contrary hypothesis, that God does not exist. Third, cast doubt on the transcendent origins of religion by showing that it has a purely natural explanation. Finally, show that we can have happy and meaningful lives without worshipping a deity and that religion, far from being a necessary prop for morality, actually produces more evil than good. The first three steps are meant to undermine the truth of religion; the last goes to its pragmatic value.

What Dawkins brings to this approach is a couple of fresh arguments—no mean achievement, considering how thoroughly these issues have been debated over the centuries—and a great deal of passion. The book fairly crackles with brio. Yet reading it can feel a little like watching a Michael Moore movie. There is lots of good, hard-hitting stuff about the imbecilities of religious fanatics and frauds of all stripes, but the tone is smug and the logic occasionally sloppy. Dawkins fans accustomed to his elegant prose might be surprised to come across such vulgarisms as "sucking up to God" and "Nur Nurny Nur Nur" (here the author, in a dubious polemical ploy, is imagining his theological adversary as a snotty playground brat). It's all in good fun when Dawkins mocks a buffoon like Pat Robertson and fundamentalist pastors like the one who created "Hell Houses" to frighten sin-prone children at Halloween. But it is less edifying when he questions the sincerity of serious thinkers who disagree with him, like the late Stephen Jay Gould, or insinuates that recipients of the million-dollar-plus Templeton Prize, awarded for work reconciling science and spirituality, are intellectually dishonest (and presumably venal to boot). In a particularly low blow, he accuses Richard Swinburne, a philosopher of religion and science at Oxford, of attempting to "justify the Holocaust" when Swinburne was struggling to square such monumental evils with the existence of a loving God. Perhaps all is fair in consciousness-raising. But Dawkins's avowed hostility can make for scattershot reasoning as well as for rhetorical excess. Moreover, in training his Darwinian guns on religion, he risks destroying a larger target than he intends.

The least satisfying part of *The God Delusion* is Dawkins's treatment of the traditional arguments for the existence of God. The "ontological argument" says that God must exist by his very nature, because he possesses all perfections, and it is more perfect to exist than not to exist. The "cosmological argument" says that the world must have an ultimate cause, and this cause could only be an eternal, godlike entity. The "design argument" appeals to special features of the universe (such as its suitability for the emergence of intelligent life), submitting that such features make it more probable than not that the universe had a purposive cosmic designer.

These, in a nutshell, are the Big Three arguments. To Dawkins, they are simply ridiculous. He dismisses the ontological argument as "infantile" and "dialectical prestidigitation" without quite identifying the defect in its logic, and he is baffled that a philosopher like Russell—"no fool," he allows—could take it seriously. He seems unaware that this argument, though medieval in origin, comes in sophisticated modern versions that are not at all easy to refute. Shirking the intellectual hard work, Dawkins prefers to move on to parodic "proofs" that he has found on the Internet, like the "Argument from Emotional Blackmail: God loves you. How could you be so heartless as not to believe in him? Therefore God exists." (For those who want to understand the weaknesses in the standard arguments for God's existence, the best source I know remains the atheist philosopher J. L. Mackie's 1982 book, *The Miracle of Theism*.)

It is doubtful that many people come to believe in God because of logical arguments, as opposed to their upbringing or having "heard a call." But such arguments, even when they fail to be conclusive, can at least give religious belief an aura of reasonableness, especially when combined with certain scientific findings. We now know that our universe burst into being some fourteen billion years ago (the theory of the big bang, as it happens, was worked out by a Belgian priest) and that its initial conditions seem to have been "fine-tuned" so that life would eventually arise. If you are not religiously inclined, you might take these as brute facts and be done with the matter. But if you think that there must be some ultimate explanation for the improbable leaping into existence of the harmonious, bio-friendly cosmos we find ourselves in, then the God hypothesis is at least rational to adhere to, isn't it?

No, it's not, says Dawkins, whereupon he brings out what he views as "the central argument of my book." At heart, this argument is an elaboration of the child's question "But, Mommy, who made God?" To posit God as the ground of all being is a nonstarter, Dawkins submits, for "any God capable of designing a universe, carefully and foresightfully tuned to lead to our evolution, must be a supremely complex and improbable entity who needs an even bigger explanation than the one he is supposed to provide." Thus the God hypothesis is "very close to being ruled out by the laws of probability."

Dawkins relies here on two premises: first, that a creator is bound to be more complex, and hence improbable, than his creation (you never, for instance, see a horseshoe making a blacksmith); and second, that to explain the improbable in terms of the more improbable is no explanation at all. Neither of these is among the "laws of probability," as he suggests. The first is hotly disputed by theologians, who insist, in a rather woolly metaphysical way, that God is the essence of simplicity. He is, after all, infinite in every respect, and therefore much easier to define than a finite thing. Dawkins, however, points out that God can't be all that simple if he is capable of, among other feats, simultaneously monitoring the thoughts of all his creatures and answering their prayers. ("Such bandwidth!" Dawkins exclaims.)

If God is indeed more complex and improbable than his creation, does that rule him out as a valid explanation for the universe? The beauty of Darwinian evolution, as Dawkins never tires of observing, is that it shows how the simple can give rise to the complex. But not all scientific explanation follows this model. In physics, for example, the law of entropy implies that for the universe as a whole order always gives way to disorder; thus, if you want to explain the present state of the universe in terms of the past, you are pretty much stuck with explaining the probable (messy) in terms of the improbable (neat). It is far from clear which explanatory model makes sense for the deepest question, the one that, Dawkins complains, his theologian friends keep harping on: Why does the universe exist at all? Darwinian processes can take you from simple to complex, but they can't take you from Nothing to Something. If there is an ultimate explanation for our contingent and perishable world, it would seemingly have to appeal to something that is both necessary and imperishable, which one might label "God." Of course, it can't be known for sure that there is such an explanation. Perhaps, as Russell thought, "the universe is just there, and that's all."

This sort of coolly speculative thinking could not be more remote from the rococo rituals of religion as it is actually practiced across the world. Why is it that all human cultures have religion if, as Dawkins believes he has proved, it rests on a delusion? Many thinkers—Marx, Freud, Durkheim—have produced natural histories of religion, arguing that it arose to serve some social or psychological function, such as, in Freud's account, the fulfillment of repressed wishes toward a father figure.

Dawkins's own attempt at a natural history is Darwinian, but not in the way you might expect. He is skeptical that religion has any survival value, contending that its cost in blood and guilt outweighs any conceivable benefits. Instead, he attributes religion to a "misfiring" of something else that is adaptively useful—namely, a child's evolved tendency to believe its parents. Religious ideas, he thinks, are virus-like "memes" that multiply by infecting the gullible brains of children. (Dawkins coined the term "meme" three decades ago to refer to bits of culture that, he holds, reproduce and compete the way genes do.) Each religion, as he sees it, is a complex of mutually compatible memes that has managed to survive a process of natural selection. ("Perhaps," he writes in his usual provocative vein, "Islam is analogous to a carnivorous gene complex, Buddhism to a herbivorous one.") Religious beliefs, according to this view, benefit neither us nor our genes; they benefit themselves.

Dawkins's gullible-child proposal is, as he concedes, just one of many Darwinian hypotheses that have been speculatively put forward to account for religion. (Another is that religion is a by-product of our genetically programmed tendency to fall in love.) Perhaps one of these hypotheses is true. If so, what would that say about the truth of religious beliefs themselves? The story Dawkins tells about religion might also be told about science or ethics. All ideas can be viewed as memes that replicate by jumping from brain to brain. Some of these ideas, Dawkins observes, spread because they are good for us, in the sense that they raise the likelihood of our genes getting into the next generation; others—like, he claims, religion—spread because normally useful parts of our minds "misfire." Ethical values, he suggests, fall into the first category. Altruism, for example, benefits our selfish genes when it is lavished on close kin who share copies of those genes or on non-kin who are in a position to return the favor. But what about pure "Good Samaritan" acts of kindness? These, Dawkins says, could be "misfirings," although, he hastens to add, misfirings of a "blessed, precious" sort, unlike the nasty religious ones.

But the objectivity of ethics is undermined by Dawkins's logic just as surely as religion is. The evolutionary biologist E. O. Wilson, in a 1985 paper written with the philosopher Michael Ruse, put the point starkly: ethics "is an illusion fobbed off on us by our genes to get us to cooperate," and "the way our biology enforces its ends is by making us think that there is an objective higher code to which we are all subject." In reducing ideas to "memes" that propagate by various kinds of "misfiring," Dawkins is, willy-nilly, courting what some have called Darwinian nihilism.

He is also hasty in dismissing the practical benefits of religion. Surveys have shown that religious people live longer (probably because they have healthier lifestyles) and feel happier (perhaps owing to the social support they get from church). Judging from birthrate patterns in the United States and Europe, they also seem to be outbreeding secular types, a definite Darwinian advantage. On the other hand, Dawkins is probably right when he says that believers are no better than atheists when it comes to behaving ethically. One classic study showed that "Jesus people" were just as likely to cheat on tests as atheists and no more likely to do altruistic volunteer work.

Oddly, Dawkins does not bother to cite such empirical evidence; instead, he relies, rather unscientifically, on his intuition. "I'm inclined to suspect," he writes, "that there are very few atheists in prison." (Even fewer Unitarians, I'd wager.) It is, however, instructive when he observes that the biblical Yahweh is an "appalling role model," sanctioning gang rape and genocide. Dawkins also deals at length with the objection, which he is evidently tired of hearing, that the arch evildoers of the last century, Hitler and Stalin, were both atheists. Hitler, he observes, "never formally renounced his Catholicism," and in the case of Stalin, a onetime Orthodox seminarian, "there is no evidence that his atheism motivated his brutality." The equally murderous Mao goes unmentioned, but perhaps it could be argued that he was a religion unto himself.

Despite the many flashes of brilliance in *The God Delusion*, Dawkins's failure to appreciate just how hard philosophical questions about religion can be makes reading it an intellectually frustrating experience. As long as there are no decisive arguments for or against the existence of God, a certain number of smart people will go on believing in him, just as smart people reflexively believe in other things for which they have no knockdown philosophical arguments, like free will, or objective values, or the existence of other minds. Dawkins asserts that "the presence or absence of a creative super-intelligence is unequivocally a scientific question." But what possible evidence could verify or falsify the God hypothesis? The doctrine that we are presided over by a loving deity has become so rounded and elastic that no earthly evil or natural disaster, it seems, can come into collision with it. Nor is it obvious what sort of event might unsettle an atheist's conviction to the contrary. Russell, when asked about this by a *Look* magazine interviewer in 1953, said he might be convinced there was a God "if I heard a voice from the sky predicting all that was going to happen to me during the next 24 hours."

Short of such a miraculous occurrence, the only thing that might resolve the matter is an experience beyond the grave—what theologians used to call, rather pompously, "eschatological verification." If the after-death options are either a beatific vision (God) or oblivion (no God), then it is poignant to think that believers will never discover that they are wrong, whereas Dawkins and fellow atheists will never discover that they are right.

As for those in between—ranging from agnostics to "spiritual" types for whom religion is not so much a metaphysical proposition as it is a way of life, illustrated by stories and enhanced by rituals—they might take consolation in the wise words of the Reverend Andrew Mackerel, the hero of Peter De Vries's 1958 comic novel *The Mackerel Plaza*: "It is the final proof of God's omnipotence that he need not exist in order to save us."

22

On Moral Sainthood

Suppose you wish to achieve excellence in life, but you have no real talent for anything in particular. You are not smart enough to be a great scientist or creative enough to be a great artist; you do not possess the native shrewdness to be a distinguished statesman or the exquisite taste (and inherited wealth) to be a legendary hedonist. Are you then doomed to mediocrity? There is a school of thought that says no. The idea is that even if you are not clever or beautiful or talented, you can still, through sheer force of will, be very, very good. You can go beyond the norms of everyday morality—being kind to people, not telling lies, giving the odd dollar to Oxfam—and devote all your energies to feeding the hungry and succoring the afflicted. In other words, you can become a moral saint.

Is it a wise idea to strive to be as good and self-sacrificing as possible? Is "moral sainthood"—a coinage of the philosopher Susan Wolf—a form of human excellence one should aspire to? Is it, indeed, something each of us has a duty to aspire to?

Philosophers, over the past couple of thousand years, have offered two reasons for aiming at the heights of moral goodness: to improve the world and to perfect one's self. These reasons do not sit together very well, because one of them is directed outward and the other is directed inward. Can you really perfect your own self by forgetting it in the service of others? There are grounds, in fact, for thinking that being as good as possible is neither good for the world nor good for one's soul.

Start with the soul part. Do those who devote their lives to the service of others tend to have beautiful, Socrates-like personalities? Are they Eros-magnetic? Many of us don't like do-gooders and are not shy about saying so. We claim to find the do-gooder earnest, meddlesome, sanctimonious, curdlingly nice. Now, a large part of this might be resentment at the fact that the do-gooder's shining example makes a mockery of our own moral pretensions. But there does seem to be something unbalanced about the soul of the perfect do-gooder. Unnaturally developed moral virtues—patience, charity in thought as well as in deed, a constant concern with alleviating the suffering of others—tend to crowd out the nonmoral virtues, like humor, intellectual curiosity, and dash.

Susan Wolf has distinguished two kinds of moral saints: the Loving Saint, whose happiness lies wholly and exclusively in helping others; and the Rational Saint, who is made happy by the same things as the rest of us—friends, family, material comforts, art, books, sports, sex—but who sacrifices his happiness out of a sense of duty. The Loving Saint, blind to so much of what life offers, has a soul that is strangely barren. The Rational Saint, who must continually suppress or deny his strongest desires, has a soul that is soured by frustration.

It's a point that has regularly been made about the humanitarians who loom largest in the public imagination: Florence Nightingale, Mahatma Gandhi, Albert Schweitzer, Mother Teresa. In Lytton Strachey's *Eminent Victorians*, Florence Nightingale was depicted as ludicrously inhuman. George Orwell scented a whiff of vanity emanating from Gandhi and concluded that "sainthood is a thing that human beings must avoid." Albert Schweitzer has been condemned as a God-playing autocrat and racist. Christopher Hitchens set off a stink bomb at the shrine of Mother Teresa, making her out to be willfully obtuse about the suffering of others. Interestingly, Dorothy Day, the heroine of the Catholic Worker Movement, has so far escaped this sort of revisionist soul blackening. Evidently, her saintliness was balanced by piquant elements left over from her bohemian-libertine past. She had redeeming vices.

Still, looking only at world-famous do-gooders leaves one open to the charge of selection bias. What happens when a fairly ordinary person takes up the cause of goodness with uncompromising zeal? That question is Nick Hornby's point of departure in his 2001 comic novel

How to Be Good. The novel's narrator is Katie Carr, a doctor in her forties who works at a depressing clinic in North London. Katie feels that she is good enough: not only does she tend the sick ("You have to be good to look at boils in the rectal area"), but she also holds nice, liberal opinions on homelessness, racism, and sexism; she reads *The Guardian* and votes Labour. Even the affair she is having as the novel opens, the first of her long marriage, is decorously conducted and quickly docked. Her husband, David, is less admirable; a sarcastic, lazy, grouchy fellow, he writes a mock-bilious newspaper column called "The Angriest Man in Holloway" and dabbles at a satirical novel "about Britain's post-Diana touchy-feely culture."

This is a marriage that has gone to rot for the usual reasons—infidelity, mutual prickliness, boredom. But just when it appears to be collapsing, David experiences a Damascene conversion. Plagued by back pain, he goes to see a flaky young healer called D. J. GoodNews, who applies warm fingertips to David's temples and sucks a "black mist" out of him. Pure benevolence rushes in to fill the vacuum; David becomes a saint. "I think everything you think. But I'm going to walk it like I talk it," he declares to his astounded wife, who suspects a brain tumor. At first, David engages in random eleemosynary acts, trying to give away the Sunday roast on the street and donating one of the kids' computers to a battered-women's shelter. Soon he undertakes a neighborhood-wide scheme to house the homeless in spare bedrooms.

The motive behind David's sudden zealotry is never made entirely clear. Though he talks a lot about the need to improve the world, he also voices a yearning for self-perfection. "I don't believe in Heaven, or anything," he says. "But I want to be the kind of person that qualifies for entry anyway." If David's soul is becoming more beautiful, the effect is lost on Katie. Her husband now strikes her as "someone from whom all trace of facetiousness, every atom of self-irony, seems to have vanished." Defective as the old David was, he at least made her laugh, which, come to think of it, was why she married him in the first place. She detects a smugness, an ostentatiousness about his good works, which puts her triumphantly in mind of a passage from one of Saint Paul's letters to the Corinthians: "Charity is not boastful, nor proud. It vaunteth not itself, is not puffed up." But when she quotes it to her husband, he reminds her that the very same passage was read at their wedding, with "love" in place of "charity"; both words, he points out, are used to translate the Latin noun *caritas*. "Love and charity share the same root word," she ponders to herself. "How is that possible, when everything in our recent history suggests that they cannot coexist, that they are antithetical, that if you put the two of them together in a sack they would bite and scratch and scream, until one of them is torn apart?"

Such reflections lead Katie to an epiphany (one that seems to bear Hornby's endorsement). The real paragon of soul beauty is not the self-sacrificing do-gooder; it is someone who leads a "rich and beautiful life," one filled with high artistic pleasures and exquisite personal affections—someone like, say, Vanessa Bell, whose biography Katie happens to be reading. And if lives of such rarefaction are no longer available to us—they are "a discontinued line," she thinks—we can at least get a tincture of them by reading Bloomsbury books and listening to tasteful chamber music. That and looking after those close to you rather than worrying about all humanity—"because life's fucking hard enough as it is"—compose the secret of how to be good.

The Hornby solution jibes agreeably with our deep-seated inclinations toward moral laziness. But is it good enough? Perhaps the do-gooders found in literature—one thinks of the ridiculous Mrs. Jellyby in *Bleak House*—are not fully representative of their counterparts in real life. In her 2015 book, *Strangers Drowning*, Larissa MacFarquhar sets out nuanced profiles of a variety of actual people who have chosen to live lives of consuming ethical commitment. One couple adopts twenty children in distress, many with disabilities; another couple founds a leprosy colony in India; a man donates a kidney to a stranger; and so forth. While the extreme altruists MacFarquhar depicts may make us uneasy by their almost masochistic devotion to succoring others, they are at least recognizably human: weird, perverse, and tetchy, no doubt, but fundamentally sane; even, on reflection, admirable. Their souls seem okay.

And some would argue that one has a duty to do as much good as possible, even if it does little for one's soul—and even if "life's fucking hard enough as it is." Suppose you see a little girl drowning in a shallow pool of water. Wouldn't you feel duty-bound to rescue her—never mind that jumping in the water means ruining your two-hundred-dollar shoes? Unfortunately, there are many children throughout the world in a situation analogous to that of the drowning girl. They are dying of starvation or of easily prevented maladies like diarrhea, and owing to the existence of a network of overseas relief organizations it is within your power to save them. Indeed, the estimated cost of rescuing one child in this way has been put at about two hundred dollars. All you have to do is call a toll-free number with your credit card in hand. Is there an important distinction between failing to do this and walking away from the drowning child? And, if there is not, can you in good conscience stop at just two hundred dollars? Don't your ethical convictions compel you to save as many children as your bank account permits?

Philosophers like Peter Singer of Princeton and Peter Unger of New York University have used such simple but intuitively powerful "rescue cases" to argue for an inconvenient conclusion: we rich Westerners should be giving away most of our money to international relief efforts. Doing so is not praiseworthy; it is required by the ethical principles we all implicitly share. Singer himself has said that he gives away a fifth of his income and wonders whether that is enough. How could it be? Even if a middle-class American family were already giving up four-fifths of its household income to Oxfam, an incremental donation of two hundred dollars would still save yet another child dying for lack of food or medical care, at the cost of relatively little distress to the giver. As for the occasional sybaritic indulgence, forget it. What's a bottle of Dom Pérignon compared with a child's life?

If that's what morality asks of me, you might say, to hell with it. And Simon Blackburn, a philosopher at Cambridge, would probably sympathize with you. In *Being Good*, his concise introduction to moral philosophy, Blackburn argues that this kind of over-demandingness threatens to undermine ethics itself. "The center of ethics must be occupied by things we can reasonably demand of each other," he writes. Our duty to help others cannot be infinite. The moral principles we adopt must not reduce us to slaves of the impersonal good. It may be praiseworthy to give away all your money in order to save starving children abroad, or to quit your Park Avenue medical practice and join Doctors Without Borders, or to invite the homeless to crash in your apartment, but it is not obligatory.

These conclusions of Blackburn's are comforting and in line with those arrived at by the narrator of Hornby's novel. Yet how are we to resist those rescue-case arguments? Blackburn does not dwell on the matter, but one line of response goes something like this. Singer is an avowed utilitarian. In its purest form, utilitarianism says that we should seek to act in a way that brings about the greatest happiness in the world. Now, one test of an ethical principle is that it be "universalizable." We ask what the world would be like if everybody acted on this principle. What if everyone devoted himself to the happiness of others? Then, on average, everyone would be less happy, because everyone would be subordinating his own happiness to the needs of others. And if everyone gave all his money to Oxfam, the resulting contraction in consumer demand would cause the world economy to collapse, leading to enormous suffering. So these principles of doing good, you might think, are collectively self-defeating.

The universalization test suggests a way of quantifying our duty to help others. Perhaps the minimum sacrifice morality requires of us is simply our "fair share": the amount that, if everyone supplied it, would result in the most happiness and the least suffering in the world. This principle makes each individual's charitable burden quite reasonable: if you have already donated a modest sum to famine relief and done a little light ladling in the soup kitchen down at your local church, you can go ahead and pop that tin of sevruga in good conscience. (Well, not entirely good conscience, because now we have to worry about the sturgeon.) Any do-gooding you engage in beyond your fair share would be what ethicists call supererogatory—commendable to perform but not blameworthy to omit.

And suppose we do go beyond the call of duty. Even if we succeed in achieving some immediate good, there is no telling what the remoter consequences of our altruism will be. All we know is that these consequences will extend far into the future, beyond the purview of our will. Owing to the contingent and chaotic nature of cause and effect, the balance of good over evil involves something like a random walk, with unforeseen reversals at every stage. (Consider the doctor who successfully delivered the fourth child of Klara and Alois Hitler after the couple's first three infants had tragically sickened and died.) It is the unknowability of the distant effects of our actions that led G. E. Moore to say, in his *Principia Ethica*, that "no sufficient reason has ever yet been found for considering one action more right or more wrong than another." Moore's conclusion, as it happens, had a liberating effect on Vanessa Bell and the Bloomsbury circle, who looked to him as their sage. Under his influence, the Bloomsburies decided that the virtues of self—the kinds of virtues that go together with a "rich and beautiful life," as the narrator of *How to Be Good* puts it—were more important than the old Victorian virtues of charity and self-sacrifice.

Once you venture outside Bloomsbury, of course, you may encounter extraordinary circumstances in which it is possible to do extraordinary good. Think of the "good Germans" who risked their lives to aid Jews during the Final Solution or of the sole member of Lieutenant Calley's platoon who conspicuously lowered his rifle instead of shooting the peasants at My Lai. What does it take to be good in such extreme situations? Skill can certainly help, as the examples of Oskar Schindler and Raoul Wallenberg and Fred Cuny show, but it is not always necessary. Is goodwill ever enough? Not unless it is combined with some other hard-to-define quality of character, which, waving our hands, we call nerve or heart. Whatever this quality is, it is one that most seemingly goodwilled people lack, which is why they become silent accomplices when great evils are being committed. In ordinary circumstances, of course, there remains the Oxfam option; yet even here there is no fixed proportion between the will to do good and the good one actually does. The widow's mite, for all the benevolence behind it, is null compared with the face-saving millions given to charities by the heartless capitalist.

So exceptional goodness always seems to require special qualifications: either great savoir faire or great bravery plus exposure to extreme circumstances that call for moral heroism. If you happen to be short on both counts, you're probably out of the running. Sainthood, it turns out, is one of those career prospects—like making millions in mail order—that seem to be open to everyone but pan out for only the fortunate few.

Beyond the necessary technical and organizational skills, achieving great altruistic feats seems to require a sort of demonic creativity. Florence Nightingale, who dedicated her life to the care of the war wounded, did an enormous amount of good (although her reforms, by reducing the human cost of war, might have made future wars more likely). But

Nightingale, as Strachey pointed out to shocked Edwardians, was not a sweet, self-abnegating angel of mercy; she was an angry, acerbic, sarcastic, and self-important woman of stern and indomitable will. She had, you could say, an artist's temperament. Evelyn Waugh once observed, "Humility is not a virtue propitious to the artist. It is often pride, emulation, avarice, malice—all the odious qualities—which drive a man to complete, elaborate, refine, destroy, renew his work until he has made something that gratifies his pride and envy and greed. And in doing so he enriches the world more than the generous and good, though he may lose his own soul in the process. That is the paradox of artistic achievement."

It may also be the paradox of altruistic achievement. If you want to be a saint, forget about being an angel.

23

Truth and Reference:
A Philosophical Feud

"Imagine the following blatantly fictional situation . . . Suppose that Gödel was not in fact the author of [Gödel's incompleteness theorem]. A man named 'Schmidt,' whose body was found in Vienna under mysterious circumstances many years ago, actually did the work in question. His friend Gödel somehow got hold of the manuscript and it was thereafter attributed to Gödel . . . So, since the man who discovered the incompleteness of arithmetic is in fact Schmidt, we, when we talk about 'Gödel,' are in fact always referring to Schmidt. But it seems to me that we are not. We simply are not . . .

"It may seem to many of you that this is a very odd example."

These words were spoken by Saul Kripke to an audience at Princeton University on January 22, 1970. Kripke, then a twenty-nine-year-old member of the Rockefeller University philosophy faculty, was in the midst of the second of three lectures that he was delivering without written text, or even notes. The lectures, which were tape-recorded and eventually published under the title *Naming and Necessity* (1980), proved to be an epoch-making event in the history of contemporary philosophy. "They stood analytic philosophy on its ear," Richard Rorty wrote in the *London Review of Books*. "Everybody was either furious, or exhilarated, or thoroughly perplexed." Kripke's lectures gave rise to what came to be called the new theory of reference, revolutionizing the way philosophers of language thought about issues of meaning and truth. They engendered hundreds of journal articles and dissertations about "possible worlds," "rigid designators," and "a pos-teriori necessity." They led to a far-reaching revival of the Aristotelian doctrine of essences. And they helped make their author, already something of a cult figure among logicians, into the very model of a modern philosophical genius—a stature *The New York Times* certified in 1977 by putting Kripke's glowering visage on the cover of its Sunday magazine.

Now imagine, if you will, the following blatantly fictional situa-tion. Suppose that Kripke was not in fact the author of the new theory of reference. A woman named Marcus—let's call her Ruth Barcan Marcus for greater verisimilitude—whose warm body can still be seen tracing out mysterious trajectories through the campus of Yale Univer-sity,* actually did the work in question. The young Kripke went to a talk she gave in 1962 containing the key ideas; almost a decade later, he presented a greatly elaborated version of them without crediting Marcus. Thereafter they were attributed to Kripke. So, because the person who discovered the new theory of reference is in fact Marcus, we, when we talk about Kripke, are in fact always referring to Marcus. Or are we?

This may seem to many of you a very odd story. Nevertheless, it is precisely the story that a philosopher by the name of Quentin Smith dared to tell a largish audience last winter† at an American Philosophi-cal Association conference in Boston. Only for Smith, a professor at Western Michigan University, the story was not blatantly fictional. It was true.

When Quentin Smith spoke to the Boston audience, it was some-thing like a philosophical version of David and Goliath—an upstart forty-two-year-old professor from a minor midwestern university at-tempting to rewrite intellectual history and take on the reputation of the man whom Robert Nozick has called "the one genius of our pro-fession." The philosophical world first got wind of Smith's charges in the fall of 1994 when the paper he was to present at the upcoming APA conference—titled "Marcus, Kripke, and the Origin of the New Theory of Reference"—was listed among the planned proceedings.

---

*Until her death, that is, in 2012.
†Nineteen ninety-four.

Before long, philosophy bulletin boards on the Internet were festooned with messages to the effect that someone was going to accuse the great Saul Kripke of plagiarism—a quite reasonable inference, given the inflammatory way the abstract for Smith's paper was worded.

The colloquium itself took place on December 28 at the Marriott hotel in Boston's Copley Place. It was not an altogether edifying spectacle. Ruth Barcan Marcus, whom Smith would be championing, did not attend. Nor did Kripke. Nor, for that matter, did most of his Princeton colleagues (their absence was interpreted by some philosophers as a token of their solidarity with Kripke, by others as a conspicuous failure to show support for him). Yet a contingent of Princeton graduate students did make their presence felt, heckling Smith ("Marcus put you up to this, didn't she?" one hostile auditor was heard to yell) and pointedly striding out of the room as he detailed the "historical misunderstanding" that led to Kripke's getting credit for ideas that, Smith claimed, were properly Marcus's. "From the point of view of the history of philosophy," Smith declared, "correcting this misunderstanding is no less important than correcting the misunderstanding in a hypothetical situation where virtually all philosophers attributed the origin of [Plato's] Theory of Forms to Plotinus."

Smith's startling claims did not go unanswered. The chosen respondent was Scott Soames, a young philosopher of language at Princeton whom some in the audience seemed to regard as a sort of philosophical hit man dispatched by Kripke's department. (In fact, he had been approached with the request to serve as commentator by the APA program committee after a couple of other philosophers declined the job.) "My task today is an unusual and not very pleasant one," Soames began, going on to rebuke Smith for his "shameful" insinuation that Kripke was guilty of intellectual theft. He heaped scorn on Smith's claim that Kripke learned the main doctrines of the new theory of reference from Marcus, misunderstood them initially, and, upon finally sorting them out in his mind, mistook them for his own—and that the rest of the philosophical profession was somehow duped in the bargain. "If there is any scandal here," Soames concluded, "it is that such a carelessly and incompetently made accusation should have been given such credence."

But that was not the end of it. Under APA rules, the colloquium speaker is allowed a reply after the commentator is finished. So Smith got up to deliver his rejoinder—which, at twenty-seven pages (not including footnotes), was nearly as long as his original paper and Soames's response combined. "I do not believe it is relevant or helpful to adopt the sort of language that Soames uses in his reply," he told the audience. "Philosophical disagreements are not solved by the disputants labeling each other's work with a variety of negative and emotive epithets; they are solved by presenting sound arguments, and I shall confine myself to presenting arguments." A smattering of applause greeted this remark.

Some way into Smith's apparently endless review of textual and philosophical minutiae, the colloquium chair, Mark Richard of Tufts, tried to cut him off. Several members of the audience objected, clamoring that he be allowed to speak on. Richard acquiesced but, in contravention of protocol, then permitted Soames a second rejoinder. ("I began to get the feeling he was acting under Soames's direction," Smith later recalled.) "If Marcus had these ideas before Kripke, how come no one said anything about it for more than twenty years?" Soames asked the audience rhetorically. "Maybe that's a question women philosophers should be asking the profession," piped up one person of gender present, causing a hush to fall briefly over the gathering.

Today, over a year later,* *l'affaire Kripke* is still alive. The colloquium papers—Smith's original, Soames's response, and Smith's counter-response—have recently been published in the philosophy journal *Synthese*. And the two adversaries are currently busy refining their briefs in another pair of papers of even greater length; Smith's latest draft is almost seventy pages. Meanwhile, such philosophical eminences as Elizabeth Anscombe, Donald Davidson, and Thomas Nagel have signed a letter to the APA asserting that "a session at a national APA meeting is not the proper forum in which to level ethical accusations against a member of our profession, even if the charges were plausibly defended." The letter, which was published in the association's quarterly proceedings, goes on to demand that the APA issue a public apology to Kripke.

---

*This account was originally published in February 1996.

The philosophical profession, it seems, has divided into several camps—defined not only by convictions about intellectual originality and propriety but also by a variety of strong feelings about Kripke the man. He is, after all, the sort of remote and brooding figure who inspires more awe than affection. His personal eccentricities have made him a subject of intense rumor-mongering. And even those who profess unstinting admiration for his intellectual achievements often complain that he has set himself up as the "policeman" of analytic philosophy, arrogantly punishing other philosophers for being derivative and stupid. And now, ironically, it is Officer Kripke himself who has come under a cloud.

Ruth Marcus declined to discuss the affair—though, to show that Smith was not "a voice in the wilderness," she did send me a dozen or so journal articles published by philosophers over the years crediting her with being an originator of the new theory of reference. By contrast, Kripke himself is quite open in ventilating his sense of hurt and exasperation. "Number one," he says, "what Smith is saying is not true, and, number two, even if it were true, the matter should have been handled more responsibly."

There is something exasperating about the matter. It is easy to tell when someone has borrowed the prose of another; one can simply look at the passages in question and see if they match, word by word. Ideas are rather trickier to identify. When a new one is discovered and put in clear, explicit form, intimations of it have a way of coming out of the woodwork of earlier texts. Was it there all along, or are we just, as it were, retrojecting? Did Oliver Heaviside really hit upon $E = mc^2$ before Einstein? Did Fermat adumbrate the fundamental theorem of calculus in advance of Newton? Can all of Freud's insights be found in *Hamlet*?

■

The daunting complexity of the ideas at stake in the Kripke/Marcus case does not make their genealogy any easier to determine. Although they mostly pertain to the philosophy of language, their deeper source is in modal logic, the formal study of the different modes of truth—necessity and possibility—that a statement can possess. First studied by Aristotle, fashionable among the medieval schoolmen, but largely neglected by their modern successors, modal logic enjoyed something of a renaissance earlier in this century, owing to the work of philosophers like C. I. Lewis and Rudolf Carnap.

In the 1940s, Ruth Barcan Marcus—then the unmarried graduate student Ruth C. Barcan—added new formal features to the apparatus of modal logic, greatly enlarging its philosophical implications. And, a decade later, the teenage prodigy Saul Kripke supplied it with something it had hitherto lacked: an interpretation, a semantics. Drawing on Leibniz's conceit that the actual world is only one in a vast collection of possible worlds—worlds where snow is green, worlds where McGovern beat Nixon—Kripke characterized a proposition as necessarily true if it holds in every possible world, and possibly true if it holds in some possible world. He then proved that modal logic was a formally "complete" system, an impressively deep result that he published in *The Journal of Symbolic Logic* in 1959 at the tender age of eighteen.

Not long thereafter, in February 1962, Kripke attended a now-legendary session at the Harvard Faculty Club. The occasion was Ruth Marcus's delivery of a paper titled "Modalities and Intensional Languages." The milieu was not a particularly clement one for the speaker, as Harvard's philosophers tended to take a dim view of the whole notion of necessity and possibility. ("Like the one whose namesake I am," Ruth Marcus later recalled, "I stood in alien corn.") This was especially true of the commentator for the paper, W. V. Quine, who, as Marcus characterized it, seemed to believe that modern modal logic was "conceived in sin"—the sin of confusing the use of a word with its mention.

Although Marcus devoted the bulk of her talk to defending modal logic against Quine's animadversions, she also used the occasion to dilate upon some ideas in the philosophy of language that she had begun to develop while working on her Ph.D. thesis in the mid-1940s, ideas concerning the relationship between a proper name and the object to which it refers. Since the beginning of the century, the received theory of proper names, conventionally attributed to Gottlob Frege and Bertrand Russell, was that every such name had associated with it a cluster of descriptions; these constituted its meaning or sense. The referent of the name was the unique object that satisfied the descriptions. According to the Frege-Russell theory, the referent of the name "Aristotle" would be the unique thing satisfying such associated descriptions as "teacher of Alexander the Great," "author of the *Metaphysics*," and so on.

If proper names are indeed descriptions in disguise, then they ought to behave like descriptions in all logical contexts—including the context of modal logic. But, as Marcus observed, they simply don't. The statement "Aristotle is Aristotle," for example, is necessarily true, whereas "Aristotle is the author of the *Metaphysics*" is merely contingent, because it is possible to imagine circumstances in which the historical Aristotle became, say, a swineherd instead of a philosopher. Such intuitions suggested to Marcus that proper names are not attached to their objects through the intervention of descriptive senses. Rather, they refer directly to their bearers, like meaningless tags. To use the older idiom of John Stuart Mill, proper names have *denotation* but no *connotation*.

The foregoing is, perforce, little more than a caricature of Marcus's actual argument, which was almost rebarbative in its complexity and abstraction. It is little wonder that in the discussion that followed on that day in Cambridge, she, Quine, and the precocious undergraduate Kripke often appeared to be talking at cross-purposes. In retrospect, however, one thing is clear: Marcus's use of modal reasoning to undermine the traditional theory of the meaning of names was a step toward the new theory of reference—a theory that emerged full-blown from Kripke's Princeton lectures a decade later. But was Marcus's work more than that?

■

This is the question that Quentin Smith began to brood on in the winter of 1990, when he received a letter from Marcus—with whom he was not personally acquainted—informing him that an allusion he had made in a published paper to the "Kripke-Donnellan theory of proper names" was not strictly accurate, given that she, too, had had a role in launching the theory. (Keith Donnellan of UCLA is another philosopher involved in the elaboration of the new theory of reference.) Smith is a boyish-looking, soft-spoken, and seemingly diffident man. He began his career as a phenomenologist but later apostatized and became an analytic philosopher. Judging from his list of publications, he is extraordinarily prolific and versatile—his book *Language and*

Time (1993) was pronounced a "masterpiece" by one reviewer, and his forthcoming works include *The Question of Ethical and Religious Meaning* and the demurely titled *Explaining the Universe.** Yet until his appearance in Boston, he was a little-known figure in his profession.

At the time he got the letter from Marcus, Smith was just starting work on a book-length history of analytic philosophy. He went back and read her "Modalities and Intensional Languages." He took a look at some of her earlier papers. From 1990 to 1994, he struggled to work out the intellectual relations between Marcus's work and Kripke's. (After her Cambridge showdown with Quine and the others, Marcus had gone on to expand her contributions to philosophical logic and to do highly influential work on the theory of belief and the nature of moral dilemmas.) Smith began to correspond regularly with Marcus but, he says, received no detailed commentary from her. Finally, he reached his conclusion: nearly all the key ideas of the new theory of reference— the very ideas with which Kripke had "stood philosophy on its ear"— were in fact due to Marcus.

That was the sensational claim that Smith unpacked before the APA conference in 1994, in a paper that, owing to what he called its "unusual nature," he was surprised the program committee accepted. After detailing six major ideas that, he maintained, had wrongly been credited to Kripke and others, he went on to suggest two reasons for the "wide misunderstanding" of the historical origins of the new theory of reference. The first was innocuous enough: despite Marcus's reputation as a pioneering figure in logic and the philosophy of language, the philosophical community had simply not paid enough attention to her early work. The second, though, was a bit unsettling. Kripke himself had failed to attribute the relevant ideas to her—and not out of malice, either, but out of obtuseness. Although he had been present at Marcus's seminal talk, the young Kripke did not really understand her ideas at the time—so, at least, Smith inferred from some of his remarks during the transcribed discussion. In the 1980 preface to *Naming and Necessity*, Kripke notes that most of the views presented therein "were formulated in about 1963–64." To Smith, this suggested that

---

*The former was published in 1997 under the title *Ethical and Religious Thought in Analytic Philosophy of Language*; the latter, as far as I can tell, was never published.

Kripke only came to grasp Marcus's arguments a year or two after she made them and that his newfound insight made it seem to him that the ideas were novel and his own. "I suspect that such instances occur fairly frequently in the history of thought and art," Smith concluded with artful blandness.

In defending Kripke against Smith's "scandalous" and "grotesquely inaccurate" brief, Scott Soames began by declaring his respect and affection for Ruth Marcus (he had been her colleague at Yale when he taught there in the late 1970s and more recently contributed to a Festschrift for her). His criticisms, he said, were aimed solely at Smith. While conceding that Marcus did deserve credit for anticipating some of the tenets of the new theory of reference, he insisted that this "in no way diminishes the seminal role of Saul Kripke." Moreover, he continued, some of the ideas that Smith attributed to Marcus—that proper names are not equivalent to descriptions, for instance—had already been formulated by other logicians, notably Frederic Fitch. This was a claim that Soames probably came to regret, for it allowed Smith to point out that Fitch was actually Marcus's adviser when she was writing her dissertation in 1943–1945, and in a paper Soames did not refer to, Fitch mentioned his indebtedness to his doctoral student for her insights.

Such palpable hits, though, are rare. For the most part, the ongoing dispute between Quentin Smith and Scott Soames over who is the real mother-father of the new theory of reference involves rather delicate philosophizing. Take the notion of rigidity. A "rigid designator" is a term that refers to the same individual in every possible world. ("Benjamin Franklin," for example, is a rigid designator, whereas "the inventor of bifocals" is not, because there are possible worlds in which someone other than Franklin can be credited with that achievement.) The phrase "rigid designator" was coined by Kripke; no one questions that. Smith, however, insists that Marcus was the one who discovered the concept (priority with words is easy, with concepts hard). Impossible! rejoins Soames: Rigid designation presupposes the more general notion of the referent of a term in a possible world, and Marcus did not have a sufficiently rich semantic framework to support such a notion. Twofold error! Smith ripostes. Not only was Marcus in possession of a semantic framework as rich as Kripke's, but such a framework is not even needed to define the concept of rigid designation. All it really takes is the basic elements of modal logic and the subjunctive mood, which, Smith adds with a flourish, are precisely the means Kripke himself used to introduce rigid designation in *Naming and Necessity*.

That is about the simplest volley one can find between these two opponents (and Soames no doubt feels he still has another shot to take). Most of the issues of attribution turn on technical arguments of such subtlety that they make, say, the epic scholarly dispute over the corrected text of *Ulysses* that took place in the 1980s look like junior-high forensics. Assemble a random jury of professional philosophers, and they probably wouldn't know what to think after listening to Smith and Soames argue their cases. And yet the allegation itself is so pointed, and so grave. If Smith is right, Kripke is diminished twice over. Not only is his reputation based on an achievement that actually belongs to another philosopher—a woman neglected by the largely male profession, no less—but he failed to realize this because he did not understand the theory at first. For a genius, the only accusation worse than intellectual theft is dimness.

∎

Happily, it is possible to get some purchase on this debate without working through all the fine points of the Smith-Soames exchange. Life is, after all, short. The simplest way to begin is by asking, just what is the new theory of reference? As a philosophical movement, it can be viewed as a reaction against several earlier currents in twentieth-century analytic philosophy. By reviving the rich metaphysical notion of "possible worlds"—and taking seriously our intuitions about them—it cocks a snook at the logical positivists, who insisted that discourse is only meaningful when it can be tested against our experience of the actual world. By freely drawing on the exotic devices of modal logic, it rejects the more down-to-earth methods of the ordinary-language philosophers, who took their inspiration from the late work of Wittgenstein.

Yet where the new theory of reference really cuts against the traditional philosophical grain is in its anti-mentalism, its refusal to make semantics depend on the contents of the minds of language users.

Meanings are not located inside the head, the theory says; they are out there in the world—the world described by science. This anti-mentalism is apparent in the claim that proper names refer to their objects directly, without the mediation of mental ideas or descriptions. But adherents of the theory don't stop there. They also argue that many common nouns—words like "gold," "tiger," and "heat"—work in the same way. Such "natural kind" terms have no definitions in the usual sense, the theory holds. What determines whether a given bit of stuff is gold, for example, is not that it is heavy, yellow, malleable, and metallic; these are merely its phenomenal properties, which might be different in another possible world. What makes it gold, rather, is its atomic structure, which, being the same in every possible world, constitutes its essence. Of course, it is a fairly recent scientific discovery that gold has the atomic number 79; before that, people talked about gold without having any concept in their heads that distinguished it from the other elements (and most people still do).

If it is not meanings in the heads of language users that connect terms like "Aristotle" and "gold" to their referents, what does do the trick? Causal chains, says the new theory of reference. The term is first applied to its object in an initial baptism—say, by an act of pointing—and is then causally passed on to others through various kinds of communicative acts: conversation, reading, and so on. Thus my present use of "Aristotle" is the latest link in a causal chain stretching backward in time (and eastward in space) to the Stagirite himself at the moment that fair name was bestowed upon him.

So the new theory of reference encompasses a slew of interrelated ideas. In an early and influential collection of articles about the new theory, titled *Naming, Necessity, and Natural Kinds* (1977), the editor Stephen Schwartz provides a dependable taxonomy. The "three main features" of the new theory, he writes—each of which "directly challenges major tenets of traditional thinking about meaning and reference"—are the following: "Proper names are rigid [they refer to the same individuals in all possible worlds]; natural kind terms are like proper names in the way they refer; and reference depends on causal chains."

How many of these three main features are due to Kripke? Smith himself allows that the second and the third, as presented by Kripke in his 1970 Princeton talks, are "genuinely new." The ideas whose provenance is being contested all fall under the first feature, the rigidity of proper names. So Soames seems to be right in asserting that while Ruth Marcus might have anticipated some of these ideas, this does not detract from Kripke's "seminal role" in the creation of the new theory of reference. Even philosophers who have had serious intellectual disagreements with Kripke tend to concur on this point. "Probably not one of these ideas is Kripke's alone, and he has never pretended otherwise," says the Rutgers philosopher Colin McGinn.* "But Kripke was the first to put them in an attractive form so that non-logicians could see their significance, and to draw out implications others didn't notice."

Yet it is hard to deny that Quentin Smith has displayed considerable brilliance, not to mention nerve, in prosecuting his case that the prime mover behind the new theory of reference was Ruth Marcus, not Saul Kripke. Indeed, it sometimes seems that Smith is too clever by half, giving Marcus credit not only for the ideas she clearly had, in more or less inchoate form, but also for all the logical consequences he has been ingenious enough to tease out of them. Smith defends his effort as a legitimate inquiry into the history of contemporary philosophy, a dispassionate presentation of philosophical arguments aimed at clearing up the genealogy of an important theory. If this inquiry has created a heated controversy, there's a simple explanation: it's just that it concerns living, active philosophers.

Smith's critics disagree. They argue that it was wrong for him to air his charges in a public forum. After all, even if he didn't directly accuse Kripke of plagiarism, he did raise delicate questions of professional ethics and intimate questions about the inner workings of Kripke's mind. "It's hard for me to remember just what my state of mind was thirty years ago," Kripke himself responds when I bring up Smith's claim.

"Sure, Ruth said in her 1962 talk that proper names were not synonymous with descriptions," he says. "A subset of the ideas I later developed were present there in a sketchy way, but there was a real paucity

---

*McGinn is now retired from the University of Miami.

of argumentation on natural language. Almost everything she was saying was already familiar to me at the time. I knew about Mill's theory of names and Russell's theory of logically proper names, and I hope that, having worked on the semantics of modal logic, I could have seen the consequences of such a position for modal logic myself. I certainly don't recall thinking, 'Wow, this is an interesting point of view; maybe I should elaborate on it,' and I doubt that any unconscious version of that thought took place."

Though Kripke chose not to respond to Smith in public, he did briefly consider taking legal action against the APA. And this past spring,* he resigned from the organization. "I remember my wife [the Princeton philosopher Margaret Gilbert] screaming when she happened to see the abstract of Smith's paper in the 1994 APA proceedings," says Kripke. "It really was worded in a libelous way. The program committee had to deal with hundreds of papers and didn't devote enough time or expertise to determining whether Smith's charges had any merit. A lawsuit has crossed my mind, but I'm reluctant to take that course because I'd have to sue the APA. And can you imagine a judge and jury trying to decide these technical matters in the philosophy of language?

"I don't think I've ever acted in bad faith," Kripke concludes. "I just try to contribute to the profession what I can. And if it keeps leading to all this backbiting, in the future I might not bother."

■

Kripke's self-defense is, in the eyes of many of his peers, a persuasive one. But it also leads them to ponder his controversial presence in the field. Is Saul Kripke an incomparable genius? A boy wonder who never fully made good? Kripke's career weighs heavily on his colleagues. And it's easy to see why.

Going back and rereading the few things Kripke has published—*Naming and Necessity*; his 1982 book, *Wittgenstein on Rules and Private Language*—one cannot help being struck by the amount of sheer pleasure they afford. For humor, lucidity, quirky inventiveness, exploratory

---

*Nineteen ninety-five.

open-mindedness, and brilliant originality, he is singularly readable among analytic philosophers. And there is his disarming candor. At one point in *Naming and Necessity*, he offhandedly remarks, "Actually sentences like 'Socrates is called "Socrates"' are very interesting and one can spend, strange as it may seem, hours talking about their analysis. I actually did, once, do that. I won't do that, however, on this occasion. (See how high the seas of language can rise. And at the lowest points, too.)"

But Kripke is no stranger to controversy. His criticisms of colleagues can be fierce. And he himself received something of a mauling for his Wittgenstein book, when the initial excitement over its appearance gave way to a backlash. Works by P.M.S. Hacker and Gordon Baker (*Scepticism, Rules, and Language*, 1984) and by Colin McGinn (*Wittgenstein on Meaning*, 1984) eventually convinced many philosophers that Kripke's interpretation was wrong. "For the first time," McGinn observes, "fallibility intruded into his life."

What's more, critics have charged Kripke with haughtily ignoring the progress that other philosophers have made on problems he's worked on. Jaakko Hintikka, a distinguished philosopher at Boston University[*] and the editor of *Synthese*, says that he decided to print Smith's and Soames's papers because "they point to a pattern in Kripke's career— a pattern that has repeated itself over and over again. Another such case occurred in 1982, when Kripke published his interpretation of Wittgenstein's 'private language argument' without any acknowledgment of the very similar work that Robert Fogelin had already published on the subject six years earlier."

Hintikka is referring to Kripke's "skeptical solution" to a paradox about rule-following raised by Wittgenstein. Fogelin, a philosopher who taught at Yale before moving to Dartmouth, had published his own interpretation of the Wittgenstein paradox in his *Wittgenstein* (1976). In the second edition of this book, Fogelin included a footnote that went on for six pages, pointing out the close parallelism between his treatment and Kripke's, arguing that nonetheless his own was more complete.

---

[*]Hintikka died in 2015 in his native Finland.

"I've no doubt that Kripke has acted in good faith," Hintikka continues. "He's not appropriating anyone else's ideas, at least consciously. He's not guilty of anything more serious than colossal naïveté and professional immaturity. The real blame in all this lies with the philosophical community—which, owing to its uncritical, romantic view of this prodigy, is far too quick to give him credit for new ideas while neglecting the contributions of others. Kripke probably got his results independently, but why should he get all the credit?"

Other philosophers agree that the "cult of genius" that has grown up around Saul Kripke might have done neither Kripke nor the profession much good. "Given the rather arid work most philosophers today do, they feel the profession needs a genius in the nineteenth-century romantic mold," comments Robert Solomon, a philosopher at the University of Texas whose books include *About Love*.* "Wittgenstein was the last genius figure we've had in our midst. He became the darling of Cambridge University, and all the students used to imitate his odd way of talking and his neurotic gestures. Now that he is gone, we need another one, and Kripke, with his own legendary eccentricities— some pleasant, some not—has been thrust into the role. He's so pampered and coddled and adored by those around him you wonder whether he can tell the difference between right and wrong. He's like an idiot savant who needs to be protected."

Solomon also raises the question of gender. "Where are all the women philosophers?" he asks. "How many cases have there been in the last two millennia where a bright male became celebrated for developing ideas first discovered by a bright female?" (When I mention this point to Kripke, he replies with vexed amusement, his raspy voice leaping into the falsetto range: "I don't think that Robert Fogelin has ever claimed to be a woman.") Several reviewers of Marcus's volume of collected papers *Modalities* (1993) have joined Smith in complaining that her early work in philosophical logic has been unjustly scanted, and have hailed her as the originator of the direct-reference idea. Perhaps, though, it is natural that most philosophers should regard a theory built around the notion of a "rigid designator" as a male thing.

---

*Solomon died in 2007.

■

By now a quarter of a century* has passed since Saul Kripke galvanized the Anglophone philosophical world with the new theory of reference—elements of which had been anticipated by Ruth Marcus a quarter of a century earlier still. The question that Smith raised is a significant one in the history of ideas. But does the new theory of reference remain a vital area of inquiry today? "Very much so," Scott Soames tells me. "The theses presented in *Naming and Necessity* have become enormously influential. They're still being extended into new areas of the philosophy of mind, like characterizing the role of belief and desire in the explanation of behavior." Others disagree. "In the late '70s and '80s, the journals were full of articles about the philosophic aspects of modal logic and the implications for truth and reference," says Barry Loewer, a philosopher at Rutgers. "Now you hardly ever see them." Robert Solomon takes a jaundiced view of the enterprise. "When people start fighting over who first got the ideas, the movement must be dead," he says. "The whole business about possible worlds and rigid designators and natural kinds is almost embarrassing in retrospect. It occupies a square millimeter in the square centimeter that constitutes a narrow conception of the philosophy of language, a tiny patch in the hectare that is philosophy." ("What does he know?" honks Kripke in response. "He's a phenomenologist.") Certainly, the philosophical outlook implicit in the new theory of reference could not be more unfashionable in the wider intellectual world at the moment. Imagine: regarding the external world as real, full of objects that have essences—essences that are disclosed not by poets or phenomenologists but by scientists!†

In the past decade, not much has been heard from Kripke. As rumor, speculation, and controversy swirl around him within the philosophical world, he seems to be little known without. When I mention him to lit-crit people, political scientists, and academics/intellectuals of other nonphilosophical stripes, the usual response is something like "Kripke? Yeah, I've heard of him. He's that guy who was on the cover

---

*Today getting onto half a century.

†Since then, fashions have shifted again. Now there is even talk in Paris of the New Objectivity.

of *The New York Times Magazine* ages ago."* People who can discuss
the philosophy of Richard Rorty in excruciating detail are unable to
identify a single idea even vaguely associated with Kripke. The name
carries little in the way of description with it; it has no "Fregean
sense" for them. Will Kripke outlast his time? Will his name survive?
The whole problem puts me in mind of a passage from *Naming and
Necessity*: "Consider Richard Feynman, to whom many of us are able
to refer. He is a leading contemporary physicist. Everyone here (I'm
sure!) can state the contents of one of Feynman's theories so as to dif-
ferentiate him from Gell-Mann. However, the man in the street, not
possessing these abilities, may still use the name 'Feynman.' When
asked he will say: well he's a physicist or something. He may not think
that this picks out anyone uniquely. I still think he uses the name 'Feynman' as a name for Feynman."

Similarly, if the new theory of reference is true, the academic on
the quad can still use the name "Kripke" to refer to Kripke—even if
all he can tell you is that "he's a philosopher or something." And if the
new theory of reference is false? If names are equivalent to descriptions? In that case, when we talk about "Kripke," we are talking about
"the inventor of the new theory of reference." And so we can be absolutely certain that Quentin Smith is mistaken, for it then becomes a
necessary truth that it's Kripke who originated the theory. Only in
some possible world, "Kripke" might be Marcus.

POSTSCRIPT

The aftermath of this controversy, and of the publication (in *Lingua
Franca*) of my account of it, is worth a word. Although Ruth Barcan
Marcus declined to talk to me while I was writing the piece, she became
a copious correspondent, both by phone and by mail, after it appeared.
She professed to be pleased with what I wrote, although she did not
like my use of the adjective "inchoate" to describe some of her ideas.

---

*Curiously, the author of the 1977 cover story about Kripke was Taylor Branch, who went on to
win a Pulitzer Prize for the first volume of his epic trilogy on Martin Luther King Jr. and the
civil rights movement.

Kripke, too, seemed to think my account was fair on the whole, judging from subsequent phone calls he made to me. Kripke and Marcus harbored lingering suspicions about each other, which they freely ventilated. It was a heady experience to be the confidant, however briefly, of these two extraordinary figures.

In the philosophical world at large, opinion on the affair pretty much lined up with the conclusion suggested by my account: that some of the key notions of the "new theory of reference" were indeed present in Marcus's work, and that her role as a precursor of this theory had been scanted, but that Kripke's enrichment of these notions, and his use of them to draw striking metaphysical implications, nevertheless constituted an original achievement, not an unjustly appropriated one. As for Quentin Smith, some of the judgments later voiced on his role in the affair were a bit harsher than mine. In *The Times Literary Supplement* of February 9, 2001, for instance, the philosopher Stephen Neale described Smith's analysis of the relative contributions of Kripke and Marcus as "confused" and "not worthy of discussion" and claimed that philosophically Smith was "out of his depth."

So the question of who deserves credit for the cluster of ideas called the new theory of reference, while not fully resolved, has at least moved closer to consensus since the affair. But what of the ideas themselves? Are they still important and vital? Has the "Kripkean revolution," now nearly half a century old, enduringly transformed the way philosophers deal with central issues in metaphysics and epistemology?

Partisans like Scott Soames insist that it has. In his weighty two-volume historical study, *Philosophical Analysis in the Twentieth Century*, published a decade after the controversy, Soames credits Kripke with restoring to professional philosophy the importance of intuitions grounded in pre-philosophical thought—like the intuition, going back to Aristotle, that things in the world have real essences. It is thanks to Kripke, Soames claims, that philosophers today can talk unblushingly about "natural kinds" and "necessary truth."

Reviewing Soames's work in the *London Review of Books*, Richard Rorty sought to deflate such claims, suggesting that Kripke's revival of essentialism would prove to be "a short-lived, reactionary fad." But that might have been wishful thinking on the part of Rorty, who had come to prefer a style of philosophy that was less professionalized and technical, more in the European tradition of being in touch with broad cultural concerns. (Rorty himself, once an imposing figure in analytic philosophy, is scarcely mentioned these days by its practitioners.) Even philosophers with reservations about the value of Kripke's contribution still concede his dominant position. "By pretty general consent, Kripke's writings (including especially *Naming and Necessity*) have had more influence on philosophy in the U.S. and U.K. than any others since the death of Wittgenstein," Jerry Fodor has observed. "Ask an expert whether there have been any philosophical geniuses in the last while, and you'll find that Kripke and Wittgenstein are the only candidates."

Today, Kripke is in his late seventies. Having retired from Princeton, he now holds the position of distinguished professor at the City University of New York, where the Saul Kripke Center has been established to tend to his intellectual legacy (which includes a vast trove of unpublished writings). In a recent informal poll to determine the "most important Anglophone philosopher, 1945–2000," Kripke placed second, behind W. V. Quine (who for some reason is never referred to as a genius).

Ruth Barcan Marcus died in 2012 at the age of ninety. Besides her more technical work in logic and the philosophy of language, she took on issues of more general interest—notably, the possibility that moral duties might come into conflict with one another. In a tribute to Marcus a few years before her death, Timothy Williamson, the Wykeham Professor of Logic at Oxford, mentioned a fact about her of which I had been unaware: that for much of her early career—from 1948, shortly after she received her doctorate from Yale, to 1963, the year after she gave the Harvard talk that the young Kripke attended—she "had no regular affiliation with a major [philosophy] department, and never applied for one. She was a wife and mother, living the life of a housewife and modal logician." Williamson went on to praise her for "ideas that are not just original, and clever, and beautiful, and fascinating, and influential, and way ahead of their time, but actually—I believe— *true*."

24

Say Anything

People have been talking bull, denying that they were talking bull, and accusing others of talking bull for ages. "Dumbe Speaker! that's a Bull," a character in a seventeenth-century English play says. "It is no Bull, to speak of a common Peace, in the place of War," a statesman from the same era declares. The word "bull," used to characterize discourse, is of uncertain origin. One venerable conjecture was that it began as a contemptuous reference to papal edicts known as bulls (from the bulla, or seal, appended to the document). Another linked it to the famously nonsensical Obadiah Bull, an Irish lawyer in London during the reign of Henry VII. It was only in the twentieth century that the use of "bull" to mean pretentious, deceitful, jejune language became semantically attached to the male of the bovine species—or, more particularly, to the excrement therefrom. Today, it is generally, albeit erroneously, thought to have arisen as a euphemistic shortening of "bullshit," a term that came into currency, dictionaries tell us, around 1915.

If "bullshit," as opposed to "bull," is a distinctively modern linguistic innovation, that could have something to do with other distinctively modern things, like advertising, public relations, political propaganda, and schools of education. "One of the most salient features of our culture is that there is so much bullshit," Harry Frankfurt, a distinguished moral philosopher who is professor emeritus at Princeton, says. The ubiquity of bullshit, he notes, is something that we have come to take for granted. Most of us are pretty confident of our ability to detect it, so we may not regard it as being all that harmful. We tend to take a more benign view of someone caught bullshitting than of someone caught lying. ("Never tell a lie when you can bullshit your way through," a father counsels his son in an Eric Ambler novel.) All of this worries Frankfurt. We cannot really know the effect that bullshit has on us, he thinks, until we have a clearer understanding of what it is. That is why we need a theory of bullshit.

Frankfurt's own effort along these lines was contained in a paper that he presented more than three decades ago at a faculty seminar at Yale. Later, that paper appeared in a journal and then in a collection of Frankfurt's writings; all the while, photocopies of it passed from fan to fan. In 2005, it was published as *On Bullshit*, a tiny book of sixty-seven spaciously printed pages that went on to become an improbable breakout success, spending half a year on the *New York Times* bestseller list.

Philosophers have a vocational bent for trying to divine the essences of things that most people never suspected had an essence, and bullshit is a case in point. Could there really be some property that all instances of bullshit possess and all non-instances lack? The question might sound ludicrous, but it is, at least in form, no different from one that philosophers ask about truth. Among the most divisive issues in philosophy today is whether there is anything important to be said about the essential nature of truth. Bullshit, by contrast, might seem to be a mere bagatelle. Yet there are parallels between the two that lead to the same perplexities.

Where do you start if you are an academic philosopher in search of the quiddity of bullshit? "So far as I am aware," Frankfurt dryly observes, "very little work has been done on this subject." He did find an earlier philosopher's attempt to analyze a similar concept under a more genteel name: humbug. Humbug, that philosopher decided, was a pretentious bit of misrepresentation that fell short of lying. (A politician talking about the importance of his religious faith comes to mind.) Frankfurt was not entirely happy with this definition. The difference between lies and bullshit, it seemed to him, was more than a matter of degree. To push the analysis in a new direction, he considers a rather peculiar anecdote about the philosopher Ludwig Wittgenstein. It was the 1930s, and Wittgenstein had gone to the hospital to visit a friend whose tonsils had just been taken out. She croaked to Wittgenstein, "I feel just like a dog that has been run over." Wittgenstein (the friend recalled) was disgusted to hear her say this. "You don't know what a dog that has been run over feels like," he snapped. Of course, Wittgenstein might simply have been joking. But Frankfurt suspects that his severity was real, not feigned. Wittgenstein was, after all, a man who devoted his life to combating what he considered pernicious forms of nonsense. What he found offensive in his friend's simile, Frankfurt guesses, was its mindlessness: "Her fault is not that she fails to get things right, but that she is not even trying."

The essence of bullshit, Frankfurt decides, is that it is produced without any concern for the truth. Bullshit needn't be false: "The bullshitter is faking things. But this does not mean that he necessarily gets them wrong." The bullshitter's fakery consists not in misrepresenting a state of affairs but in concealing his own indifference to the truth of what he says. The liar, by contrast, is concerned with the truth, in a perverse sort of fashion: he wants to lead us away from it. As Frankfurt sees it, the liar and the truth teller are playing on opposite sides of the same game, a game defined by the authority of truth. The bullshitter opts out of this game altogether. Unlike the liar and the truth teller, he is not guided in what he says by his beliefs about the way things are. And that, Frankfurt says, is what makes bullshit so dangerous: it unfits a person for telling the truth.

Frankfurt's account of bullshit is doubly remarkable. Not only does he define it in a novel way that distinguishes it from lying; he also uses this definition to establish a powerful claim: "Bullshit is a greater enemy of truth than lies are." If this is true, we ought to be tougher on someone caught bullshitting than we are on someone caught lying. Unlike the bullshitter, the liar at least cares about the truth: he is anxious that his assertions be correlated with the truth, albeit negatively; so he is concerned to get the picture right, if only to invert that picture.

But isn't this account a little too flattering to the liar? In theory, of course, there could be liars who are motivated by sheer love of deception. This type was identified by Saint Augustine in his treatise *On Lying*. Someone who tells a lie as a means to some other goal tells it "unwillingly," Augustine says. The pure liar, by contrast, "takes delight in lying, rejoicing in the falsehood itself." But such liars are exceedingly rare, as Frankfurt concedes. Not even Iago had that purity of heart.

Ordinary tellers of lies simply aren't principled adversaries of the truth. Suppose an unscrupulous used-car salesman is showing you a car. He tells you that it was owned by a little old lady who drove it only on Sundays. The engine's in great shape, he says, and it runs beautifully. Now, if he knows all this to be false, he's a liar. But is his goal to get you to believe the opposite of the truth? No, it's to get you to buy the car. If the things he was saying happened to be true, he'd still say them. He'd say them even if he had no idea who the car's previous owner was or what condition the engine was in.

How do we square this case with Frankfurt's hard distinction between lying and bullshitting? Frankfurt would say that this used-car salesman is a liar only by accident. Even if he happens to know the truth, he decides what he's going to say without caring what it is. But then surely almost every liar is, at heart, a bullshitter. Both the liar and the bullshitter typically have a goal. It may be to sell a product, to get votes, to keep a spouse from walking out of a marriage in the wake of embarrassing revelations, to make someone feel good about himself, to mislead Nazis who are looking for Jews. The alliance the liar strikes with untruth is one of convenience, to be abandoned the moment it ceases to serve this goal.

The porousness of Frankfurt's theoretical boundary between lies and bullshit is apparent in Laura Penny's 2005 book, *Your Call Is Important to Us: The Truth About Bullshit*. The author, a young Canadian college teacher and former union organizer, begins by saluting Frankfurt's "subtle and useful" distinction: "The liar still cares about the truth. The bullshitter is unburdened by such concerns." She then proceeds to apply the term "bullshit" to every kind of trickery by which powerful, moneyed interests attempt to gull the public. "Most of what passes for news," Penny submits, "is bullshit"; so is the language employed by lawyers and insurance men; so is the use of rock songs in ads. She even stretches the rubric to apply to things as well as to words. "The new product that will change your life is probably just more cheap, plastic bullshit," she writes. At times, despite her nod to Frankfurt, Penny appears to equate bullshit with deliberate deceit: "Never in the history of mankind have so many people uttered statements they know to be untrue." But then she says that George W. Bush ("a world-historical bullshitter") and his circle "distinguish themselves by believing their own bullshit," which suggests that they themselves are deluded. One wonders whether she would extend the same backhanded tribute to Donald Trump.

Frankfurt concedes that in popular usage "bullshit" is employed as a "generic term of abuse, with no very specific literal meaning." What he wanted to do, he says, was to get to the essence of the thing in question. But does bullshit have a single essence? In a paper titled "Deeper into Bullshit," the Oxford philosopher G. A. Cohen protested that Frankfurt excludes an entire category of bullshit: the kind that appears in academic works. If the bullshit of ordinary life arises from indifference to truth, Cohen says, the bullshit of the academy arises from indifference to meaning. It may be perfectly sincere, but it is nevertheless nonsensical. Cohen, who was a specialist in Marxism, complains of having been grossly victimized by this kind of bullshit as a young man back in the 1960s, when he did a lot of reading in the French school of Marxism inspired by Louis Althusser. So traumatized was he by his struggle to make some sense of these defiantly obscure texts that he went on to found, at the end of the 1970s, a Marxist discussion group that took as its motto *Marxismus sine stercore tauri*—"Marxism without the shit of the bull."

Anyone familiar with the varieties of "theory" that have made their way from the Left Bank of Paris into American English departments will be able to multiply examples of the higher bullshit ad libitum. But it would be hasty to dismiss all unclear discourse as bullshit. Cohen adduces a more precise criterion: the discourse must be not only unclear but unclarifiable. That is, bullshit is the obscure that cannot be rendered unobscure. How would one defend philosophers like Hegel and Heidegger from the charge that their writings are bullshit? Not, Cohen says, by showing that they cared about the truth (which would be enough to get them off the hook if they were charged with being bullshitters under Frankfurt's definition). Rather, one would try to show that their writings actually made some sense. And how could one prove the opposite, that a given statement is hopelessly unclear and hence bullshit? One proposed test is to add a "not" to the statement and see if that makes any difference to its plausibility. If it doesn't, that statement is bullshit. As it happens, Heidegger once came very close to doing this himself. In the fourth edition of his treatise *What Is Metaphysics?*

(1943), he asserted, "Being can indeed be without beings." In the fifth edition (1949), this sentence became "Being never is without beings."

Frankfurt acknowledges the higher bullshit as a distinctive variety, but he doesn't think it's very dangerous compared with the sort of bullshit that he is concerned about. While genuinely meaningless discourse may be "infuriating," he says, it is unlikely to be taken seriously for long, even in the academic world. The sort of bullshit that involves indifference to veracity is far more insidious, Frankfurt claims, because the "conduct of civilized life, and the vitality of the institutions that are indispensable to it, depend very fundamentally on respect for the distinction between the true and the false."

How evil is the bullshitter? That depends on how valuable truthfulness is. When Frankfurt observes that truthfulness is crucial in maintaining the sense of trust on which social cooperation depends, he's appealing to truth's instrumental value. Whether it has any value in itself, however, is a separate question. To take an analogy, suppose a well-functioning society depends on the belief in God, whether or not God actually exists. Someone of subversive inclinations might question the existence of God without worrying too much about the effect that might have on public morals. And the same attitude is possible toward truth. As the philosopher Bernard Williams observed in a book published in 2002, not long before his death, a suspicion of truth has been a prominent current in modern thought. It was something that Williams found lamentable. "If you do not really believe in the existence of truth," he asked, "what is the passion for truthfulness a passion for?"

The idea of questioning the existence of truth might seem bizarre. No sane person doubts that the distinction between true and false is sharp enough when it comes to statements like "Saddam had WMDs" or "Carbon emissions contribute to climate change" or "The cat is on the mat." But when it comes to more interesting propositions—assertions of right and wrong, judgments of beauty, grand historical narratives, talk about possibilities, scientific statements about unobservable entities—the objectivity of truth becomes harder to defend. "Deniers" of truth (as Williams called them) insist that each of us is trapped in his own point of view; we make up stories about the world and, in an exercise of power, try to impose them on others.

The battle lines between deniers and defenders of absolute truth are strangely drawn. On the pro-truth side, one finds the pope emeritus Benedict XVI, who staunchly maintained that moral truths correspond to divine commands and railed against what he (oddly) calls the "dictatorship of relativism." On the "anything goes" side, one finds the member of the George W. Bush administration who mocked the idea of objective evidence by declaring, "We're an empire now, and when we act, we create our own reality." Among philosophers, Continental post-structuralists like Jean Baudrillard and Jacques Derrida tend to be arrayed on the anti-truth side. One might expect their hardheaded counterparts in Britain and the United States—practitioners of what is called analytic philosophy—to be firmly in the pro-truth camp. And yet, as Simon Blackburn observes in his 2005 book, *Truth: A Guide*, the "brand-name" Anglophone philosophers of the past fifty years—Wittgenstein, W. V. Quine, Thomas Kuhn, Donald Davidson, Richard Rorty—have developed powerful arguments that seem to undermine the commonsense notion of truth as agreement with reality. Indeed, Blackburn says, "almost all the trends in the last generation of serious philosophy lent aid and comfort to the 'anything goes' climate"— the very climate that, Harry Frankfurt argues, has encouraged the proliferation of bullshit.

Blackburn, who is himself a professor of philosophy at Cambridge University, wants to rally the pro-truth forces. But he is also concerned to give the other side its due. In *Truth*, he scrupulously considers the many forms that the case against truth has taken, going back as far as the ancient Greek philosopher Protagoras, whose famous saying "Man is the measure of all things" was seized upon by Socrates as an expression of dangerous relativism. In its simplest form, relativism is easy to refute. Take the version of it that Richard Rorty once lightheartedly offered: "Truth is what your contemporaries let you get away with." The problem is that contemporary Americans and Europeans won't let you get away with that characterization of truth; so, by its own standard, it cannot be true. (Sidney Morgenbesser's gripe about pragmatism— which, broadly speaking, equates truth with usefulness—was in the same spirit: "It's all very well in theory, but it doesn't work in practice.") Then there is the often heard complaint that the whole truth will always elude us. Fair enough, Blackburn says, but partial truths can still be perfectly objective. He quotes Clemenceau's riposte to skeptics who asked what future historians would say about World War I: "They will not say that Belgium invaded Germany."

If relativism needed a bumper-sticker slogan, it would be Nietzsche's dictum "There are no facts, only interpretations." Nietzsche was inclined to write as if truth were manufactured rather than discovered, a matter of manipulating others into sharing our beliefs rather than getting those beliefs to "agree with reality." In another of his formulations, "Truths are illusions that we have forgotten are illusions." If that's the case, then it is hard to regard the bullshitter, who does not care about truth, as all that villainous. Perhaps, to paraphrase Nietzsche, truth is merely bullshit that has lost its stench. Blackburn has ambivalent feelings about Nietzsche, who, were it not for his "extraordinary acuteness," would qualify as "the pub bore of philosophy." Yet, he observes, at the moment Nietzsche is the most influential of the great philosophers, not to mention the "patron saint of postmodernism," so he must be grappled with. One of Nietzsche's more notorious doctrines is perspectivism—the idea that we are condemned to see the world from a partial and distorted perspective, one defined by our interests and values. Whether this doctrine led Nietzsche to a denial of truth is debatable: in his mature writings, at least, his scorn is directed at the idea of metaphysical truth, not at the scientific and historical varieties. Nevertheless, Blackburn accuses Nietzsche of sloppy thinking. There is no reason, he says, to assume that we are forever trapped in a single perspective or that different perspectives cannot be ranked according to accuracy. And if we can move from one perspective to another, what is to prevent us from conjoining our partial views into a reasonably objective picture of the world?

In the American academy, Richard Rorty was probably the most prominent "truth denier" in recent times. (He died in 2007.) What made Rorty so formidable is the clarity and eloquence of his case against truth and, by implication, against the Western philosophical tradition. Our minds do not "mirror" the world, he argued. The idea that we could somehow stand outside our own skins and survey the relationship between our thoughts and reality is a delusion. We can have a theory about that relationship, but the theory is just another thought. Language is an adaptation, and the words we use are tools. There are many competing vocabularies for talking about the world, some more useful than others, given human needs and interests. None of them, however, correspond to the Way Things Really Are. Inquiry is a process of reaching a consensus on the best way of coping with the world, and "truth" is just a compliment we pay to the result. Rorty was fond of quoting the American pragmatist John Dewey to the effect that the search for truth is merely part of the search for happiness. He also liked to cite Nietzsche's observation that truth is a surrogate for God. Asking of someone, "Does he love the truth?" Rorty said, is like asking, "Is he saved?" In our moral reasoning, we no longer worry about whether our conclusions correspond to the divine will; so in the rest of our inquiry we ought to stop worrying about whether our conclusions correspond to a mind-independent reality.

Do Rorty's arguments offer aid and comfort to bullshitters? Blackburn thinks so. Creating a consensus among their peers is something that hardworking laboratory scientists try to do. But it is also what creationists and Holocaust deniers do. Rorty was careful to insist that even though the distinction between truth and consensus is untenable, we can distinguish between "frivolous" and "serious." Some people are "serious, decent, and trustworthy"; others are "unconversable, incurious, and self-absorbed." Blackburn thinks that the only way to make this distinction is by reference to the truth: serious people care about it, whereas frivolous people do not. Yet there is another possibility that can be extrapolated from Rorty's writings: serious people care not only about producing agreement but also about justifying their methods for producing agreement. (This is, for example, something that astrophysicists do but astrologers don't.) That, and not an allegiance to some transcendental notion of truth, is the Rortian criterion that distinguishes serious inquirers from bullshitters.

Pragmatists and perspectivists are not the only enemies Blackburn considers, though, and much of his book is taken up with contemporary arguments turning on subversive-sounding expressions like "holism," "incommensurability," and the "Myth of the Given." Take the last of these. Our knowledge of the world, it seems reasonable to suppose, is founded on causal interactions between us and the things in it. The molecules and photons impinging on our bodies produce sensations; these sensations give rise to basic beliefs—like "I am seeing red now"—which serve as evidence for higher-level propositions about the world. The tricky part of this scheme is the connection between sensa-tion and belief. How does a "raw feel" transform itself into something like a proposition? As William James wrote, "A sensation is rather like a client who has given his case to a lawyer and then has passively to listen in the courtroom to whatever account of his affairs, pleasant or unpleasant, the lawyer finds it most expedient to give." The idea that a sensation can enter directly into the process of reasoning has become known as the Myth of the Given. The American philosopher Donald Davidson, whose influence in the Anglophone philosophical world was unsurpassed, put the point succinctly: "Nothing can count as a reason for holding a belief except another belief."

This line of thought, as Blackburn observes, threatens to cut off all contact between knowledge and the world. If beliefs can be checked only against other beliefs, then the sole criterion for a set of beliefs' being true is that they form a coherent web: a picture of knowledge known as holism. And different people interacting with the causal flux that is the world might well find themselves with distinct but equally coherent webs of belief—a possibility known as incommensu-rability. In such circumstances, who is to say what is truth and what is bullshit? But Blackburn will have none of this. The slogan "Nothing can count as a reason for holding a belief except another belief" can't be right, he claims. After all, if "John comes in and gets a good doggy whiff, he acquires a reason for believing that Rover is in the house. If Mary looks in the fridge and sees the butter, she acquires a reason for believing that there is butter in the fridge."

Not so fast, a Davidsonian might reply. Sensations do not come la-beled as "doggy whiffs" or "butter sightings"; such descriptions imply a good deal of prior concept formation. What gives John a reason to believe that Rover is in the house is indeed another belief: that what he is smelling falls under the category of "doggy whiff." Blackburn is obviously right in maintaining that such beliefs arise from causal in-teraction with the world and not just from voices in our heads. But justifying those beliefs—determining whether we are doing well or badly in forming them—can be a matter only of squaring them with other beliefs. Derrida was not entirely bullshitting when he said, "Il n'y a pas de hors-texte" (There is nothing outside the text).

Although Blackburn concludes that objective truth can and must

survive the assaults of its critics, he himself has been forced to dimin-
ish that which he would defend. He and his allies, one might think,
should be willing to give some sort of answer to the question that "jesting
Pilate" put to Jesus: What is truth? The most obvious answer, that
truth is correspondence to the facts, founders on the difficulty of saying just what form this "correspondence" is supposed to take and what
"facts" could possibly be other than truths themselves. Indeed, about
the only thing that everyone can agree on is that each statement sup-
plies its own conditions for being true. The statement "Snow is white"
is true if and only if snow is white; the statement "The death penalty
is wrong" is true if and only if the death penalty is wrong; and so forth.
As far as Blackburn is concerned, any attempt to go beyond this simple
observation by trying to mount a general theory of what makes things
true or false is wrongheaded. That makes him, to use his own term, a
"minimalist" about truth. By reducing truth to something "small and
modest," Blackburn hopes to induce its enemies to call off their siege.

The problem with this minimalist strategy is that it leaves us with
little to care about. If truth necessarily eludes our theoretical grasp, then
how do we know that it has any value, let alone that it is an absolute
good? Why should we worry about whether our beliefs deserve to be
called "true"? Deep down, we might prefer to believe whatever helps
us achieve our ends and enables us to flourish, regardless of whether it
is true. We may be happier believing in God even if there is no God.
We may be happier thinking that we are really good at what we do
even if that is a delusion. (The people with the truest understanding of
their own abilities, research suggests, tend to be depressives.)

So truth may not be an absolute good. And even its instrumental
value might be overrated. Still, one thing can be said in its favor: it
has an aesthetic edge on bullshit. Most bullshit is ugly. When it takes
the form of political propaganda, management-speak, or PR, it is
riddled with euphemism, cliché, fake folksiness, false emotion, and high-
sounding abstractions. Yet bullshit doesn't have to be ugly. Indeed,
much of what we call poetry consists of trite or false ideas dressed up
in sublime language—ideas like "beauty is truth, truth beauty," which
is beautiful but untrue. (Oscar Wilde, in his dialogue "The Decay of
Lying," suggests that the proper aim of art is "the telling of beautiful
untrue things.")

The aesthetic dimension of bullshit is largely ignored in Frankfurt's essay. Yet he does concede that bullshitting can involve an element of artistry; it offers opportunities for "improvisation, color, and imaginative play." The problem is that most bullshitting is done from an ulterior motive. When the aim is the selling of a product or the manipulation of an electorate, the outcome is likely to be a ghastly abuse of language.

When it is done for its own sake, however, something delightful just might result. The paradigm here is Sir John Falstaff, the greatest comic genius literature has to offer. In a dark Shakespearean world of politics and war, statecraft and treachery, Falstaff stands as a beacon of freedom. He refuses to be enslaved by anything that might interfere with his ease—least of all by the authority of truth. The fat knight's beautiful strain of bullshit makes him both a wit and the cause of wit in others. He is an enemy of priggishness and respectability, a model of merriment and comradeship—far better company than the dour Wittgenstein. We should by all means be severe in dealing with bullshitters of the political, the commercial, and the academic varieties. But let's not banish sweet Jack.

Further Reading

1. WHEN EINSTEIN WALKED WITH GÖDEL

John S. Rigden, *Einstein 1905: The Standard of Greatness* (Harvard, 2005).

Rebecca Goldstein, *Incompleteness: The Proof and Paradox of Kurt Gödel* (Norton, 2005).

Palle Yourgrau, *A World Without Time: The Forgotten Legacy of Gödel and Einstein* (Allen Lane, 2005).

2. TIME—THE GRAND ILLUSION?

Paul Davies, *About Time: Einstein's Unfinished Revolution* (Simon & Schuster, 1995).

J. Richard Gott, *Time Travel in Einstein's Universe: The Physical Possibilities of Travel Through Time* (Houghton Mifflin, 2001).

Huw Price, *Time's Arrow and Archimedes' Point: New Directions for the Physics of Time* (Oxford, 1996).

3. NUMBERS GUY: THE NEUROSCIENCE OF MATH

Stanislas Dehaene, *The Number Sense: How the Mind Creates Mathematics*, rev. ed. (Oxford, 2011).

Stanislas Dehaene, *Consciousness and the Brain: Deciphering How the Brain Codes Our Thoughts* (Viking, 2014).

Brian Butterworth, *What Counts: How Every Brain Is Hardwired for Math* (Free Press, 1999).

4. THE RIEMANN ZETA CONJECTURE AND THE LAUGHTER OF THE PRIMES

Karl Sabbagh, *The Riemann Hypothesis: The Greatest Unsolved Problem in Mathematics* (Farrar, Straus and Giroux, 2003).

Marcus du Sautoy, *The Music of the Primes: Searching to Solve the Greatest Mystery in Mathematics* (Harper, 2003).

V. S. Ramachandran and Sandra Blakeslee, *Phantoms in the Brain: Probing the Mysteries of the Human Mind* (Morrow, 1998).

5. SIR FRANCIS GALTON, THE FATHER OF STATISTICS ...
AND EUGENICS

Martin Brookes, *Extreme Measures: The Dark Visions and Bright Ideas of Francis Galton* (Bloomsbury, 2004).

Daniel J. Kevles, *In the Name of Eugenics: Genetics and the Uses of Human Heredity* (Knopf, 1985).

Stephen M. Stigler, *The History of Statistics: The Measurement of Uncertainty Before 1900* (Belknap, 1986).

6. A MATHEMATICAL ROMANCE

Edward Frenkel, *Love and Math: The Heart of Hidden Reality* (Basic, 2013).

E. T. Bell, *Men of Mathematics* (repr., Touchstone, 1986).

7. THE AVATARS OF HIGHER MATHEMATICS

G. H. Hardy, *A Mathematician's Apology* (Cambridge, 1940).

Michael Harris, *Mathematics Without Apologies: Portrait of a Problematic Vocation* (Princeton, 2015).

8. BENOIT MANDELBROT AND THE DISCOVERY OF FRACTALS

Benoit Mandelbrot, *The Fractalist: Memoir of a Scientific Maverick* (Pantheon, 2012).

Benoit Mandelbrot and Richard L. Hudson, *The (Mis)behavior of Markets: A Fractal View of Financial Turbulence* (Basic, 2006).

9. GEOMETRICAL CREATURES

Edwin A. Abbott, *The Annotated Flatland: A Romance of Many Dimensions*, with an introduction and notes by Ian Stewart (Perseus, 2002).

Lawrence M. Krauss, *Hiding in the Mirror: The Quest for Alternate Realities, from Plato to String Theory* (Viking, 2005).

10. A COMEDY OF COLORS

Robin Wilson, *Four Colors Suffice: How the Map Problem Was Solved* (Princeton, 2003).

Ian Stewart, *Visions of Infinity: The Great Mathematical Problems* (Basic, 2013).

11. INFINITE VISIONS: GEORG CANTOR V. DAVID FOSTER WALLACE

David Foster Wallace, *Everything and More: A Compact History of* ∞ (Norton, 2003).

Shaughan Lavine, *Understanding the Infinite* (Harvard, 1994).

12. WORSHIPPING INFINITY: WHY THE RUSSIANS DO AND THE FRENCH DON'T

Loren Graham and Jean-Michel Kantor, *Naming Infinity: A True Story of Religious Mysticism and Mathematical Creativity* (Belknap, 2009).

Rudy Rucker, *Infinity and the Mind: The Science and Philosophy of the Infinite* (Princeton, 1995).

13. THE DANGEROUS IDEA OF THE INFINITESIMAL

Amir Alexander, *Infinitesimal: How a Dangerous Mathematical Theory Shaped the Modern World* (Scientific American / Farrar, Straus and Giroux, 2014).

Michel Blay, *Reasoning with the Infinite: From the Closed World to the Mathematical Universe*, trans. M. B. DeBevoise (Chicago, 1998).

Joseph Warren Dauben, *Abraham Robinson: The Creation of Nonstandard Analysis, a Personal and Mathematical Odyssey* (Princeton, 1995).

14. THE ADA PERPLEX: WAS BYRON'S DAUGHTER THE FIRST CODER?

Dorothy Stein, *Ada: A Life and Legacy* (MIT, 1987).

Benjamin Woolley, *The Bride of Science: Romance, Reason, and Byron's Daughter* (McGraw-Hill, 1999).

15. ALAN TURING IN LIFE, LOGIC, AND DEATH

Andrew Hodges, *Alan Turing: The Enigma* (Walker, 2000).

David Leavitt, *The Man Who Knew Too Much: Alan Turing and the Invention of the Computer* (Norton, 2006).

Martin Davis, *Engines of Logic: Mathematics and the Origin of the Computer* (Norton, 2000).

16. DR. STRANGELOVE MAKES A THINKING MACHINE

George Dyson, *Turing's Cathedral: The Origins of the Digital Universe* (Pantheon, 2012).

Norman MacRae, *John von Neumann: The Scientific Genius Who Pioneered the Modern Computer, Game Theory, Nuclear Deterrence, and Much More* (Pantheon, 1992).

17. SMARTER, HAPPIER, MORE PRODUCTIVE

Nicholas Carr, *The Shallows: What the Internet Is Doing to Our Brains* (Norton, 2010).

Steven Johnson, *Everything Bad Is Good for You: How Today's Popular Culture Is Actually Making Us Smarter* (Riverhead, 2006).

Gary Marcus, *Kluge: The Haphazard Construction of the Human Mind* (Houghton Mifflin, 2008).

18. THE STRING THEORY WARS: IS BEAUTY TRUTH?

Brian Greene, *The Elegant Universe: Superstrings, Hidden Dimensions, and the Quest for the Ultimate Theory* (Norton, 1999).

Lee Smolin, *The Trouble with Physics: The Rise of String Theory, the Fall of a Science, and What Comes Next* (Houghton Mifflin, 2006).

Peter Woit, *Not Even Wrong: The Failure of String Theory and the Search for Unity in Physical Law* (Basic, 2006).

19. EINSTEIN, "SPOOKY ACTION," AND THE REALITY OF SPACE

George Musser, *Spooky Action at a Distance: The Phenomenon That Reimagines Space and Time—and What It Means for Black Holes, the Big Bang, and Theories of Everything* (Scientific American / Farrar, Straus and Giroux, 2015).

Tim Maudlin, *Quantum Non-Locality and Relativity: Metaphysical Intimations of Modern Physics*, 3rd ed. (Wiley-Blackwell, 2011).

20. HOW WILL THE UNIVERSE END?

Steven Weinberg, *The First Three Minutes: A Modern View of the Origin of the Universe* (Basic, 1977).

Sean Carroll, *From Eternity to Here: The Quest for the Ultimate Theory of Time* (Plume, 2010).

Paul Davies, *The Last Three Minutes: Conjectures About the Ultimate Fate of the Universe* (Basic, 1994).

21. DAWKINS AND THE DEITY

Richard Dawkins, *The God Delusion* (Houghton Mifflin, 2006).

J. L. Mackie, *The Miracle of Theism: Arguments for and Against the Existence of God* (Oxford, 1982).

Richard Swinburne, *Is There a God?* (Oxford, 1996).

22. ON MORAL SAINTHOOD

Nick Hornby, *How to Be Good* (Riverhead, 2001).

Simon Blackburn, *Being Good: A Short Introduction to Ethics* (Oxford, 2001).
Larissa MacFarquhar, *Strangers Drowning: Grappling with Impossible Idealism, Drastic Choices, and the Overpowering Urge to Help* (Penguin, 2015).

23. TRUTH AND REFERENCE: A PHILOSOPHICAL FEUD
Ruth Barcan Marcus, *Modalities: Philosophical Essays* (Oxford, 1993).
Saul Kripke, *Naming and Necessity* (Wiley-Blackwell, 1991).
A. J. Ayer, *Philosophy in the Twentieth Century* (Vintage, 1982).

24. SAY ANYTHING
Harry G. Frankfurt, *On Bullshit* (Princeton, 2005).
Simon Blackburn, *Truth: A Guide* (Oxford, 2005).
Richard Rorty, *Truth and Progress: Philosophical Papers*, vol. 3 (Cambridge, 1998).

Acknowledgments

The longer essays in this volume previously appeared, in somewhat different form, in the following publications: "A Mathematical Romance," "The Avatars of Higher Mathematics," "Benoit Mandelbrot and the Discovery of Fractals," "Geometrical Creatures," "A Comedy of Colors," "The Dangerous Idea of the Infinitesimal," "Dr. Strangelove Makes a Thinking Machine," and "Einstein, 'Spooky Action,' and the Reality of Space" in *The New York Review of Books*; "When Einstein Walked with Gödel," "Numbers Guy: The Neuroscience of Math," "Sir Francis Galton, the Father of Statistics . . . and Eugenics," "Infinite Visions: Georg Cantor v. David Foster Wallace," "The Ada Perplex: Was Byron's Daughter the First Coder?," "Alan Turing in Life, Logic, and Death," "The String Theory Wars: Is Beauty Truth?," "On Moral Sainthood," and "Say Anything" in *The New Yorker*; "Worshipping Infinity: Why the Russians Do and the French Don't" and "Smarter, Happier, More Productive" in the *London Review of Books*; "Dawkins and the Deity" in *The New York Times Book Review*; "How Will the Universe End?" in *Slate*; "Time—the Grand Illusion?" in *Lapham's Quarterly*; "Truth and Reference: A Philosophical Feud" in *Lingua Franca*; and "The Riemann Zeta Conjecture and the Laughter of the Primes" in *Year Million: Science at the Far Edge of Knowledge*, edited by Damien Broderick (Atlas, 2008). The shorter essays appeared in *Lingua Franca*, with the exception of "Newcomb's Problem and the Paradox of Choice" and "Can't Anyone Get Heisenberg Right?," which appeared in *Slate*; "Death: Bad?," which appeared in *The New York Times Book Review*; and "The Mind of a Rock," which appeared in *The New York Times Magazine*.

I am grateful to the following editors: the late Robert Silvers, at *The New York Review of Books*; Henry Finder and Leo Carey, at *The New Yorker*; Mary-Kay Wilmers and Paul Myerscough, at the *London Review of Books*; Jacob Weisberg, Meghan O'Rourke, and Jack Shafer, at *Slate*; Sam Tanenhaus and Jenny Schuessler, at *The New York Times Book Review*; Lewis Lapham and Kelly Burdick, at *Lapham's Quarterly*; and Alex Star, at *The New York Times Magazine* and *Lingua Franca*.

Index